D0761990

Milking
the Public

Political Scandals
of the Dairy
Lobby from L.B.J.
to Jimmy Carter

Milking the Public

Political Scandals of the Dairy Lobby from L.B.J. to Jimmy Carter

Michael McMenamin
and
Walter McNamara

Nelson-Hall 𝗻𝗵 Chicago

Library of Congress Cataloging in Publication Data

McMenamin, Michael.
  Milking the public.

  Includes index.
    1. Milk trade—United States—Political
aspects. 2. Milk—Prices—United States—
Political aspects. 3. Corruption in politics)—
United States. I. McNamara, Walter, joint author.
II. Title.
HD9282.U4M25   328.73′078   80–11546
ISBN 0–88229–552–7

Manufactured in the United States of America

10   9   8   7   6   5   4   3   2   1

To Carol and Mardy

Contents

Acknowledgments

Our special thanks to the following persons for their kind help and assistance:

Paul W. Walter, the senior partner in our law firm, who first introduced us to the milk business and whose many years of experience, which span the modern history of the dairy lobby, were generously available for our use.

Robert Poole, Jr., the editor of *Reason* magazine, whose suggestion in 1975 for an article on milk and the corporate state eventually grew into this book.

Art Hoppe, the syndicated humor columnist, who agreed to the reprinting of his article "How the Milky Way Turned to Gold," which conveys about the dairy lobby in a few hundred words what it took us an entire book to do.

Diane Connelly, formerly of *Reader's Digest,* who worked with us on nearly every aspect of the book—research, interviews, editing, and re-writing—and for whose excellent and perceptive contributions we are particularly grateful.

Dr. Roland W. Bartlett, Professor of Agricultural Economics, *Emeritus,* at the University of Illinois, Jackson E. Betts, a former congressman from Ohio, and our fraternity brother, William Dobrovir, a Washington, D.C. attorney, Al Nixon, Executive Director of the National Association for Milk Marketing Reform, and Frank Wright, Managing Editor of the *Minneapolis Tribune,* all of whom graciously read portions of our manuscript and offered their comments and suggestions.

Linda English, for whose many, many hours of deciphering, rearranging, typing and retyping the manuscript early in the morning and on weekends we are most appreciative, and Doris Hribar, for her excellent help and assistance in a similar capacity.

There are many others who helped us in the preparation of this book—in the government, in the dairy industry, in our law firm and our families—and, while space does not permit us to thank them all individually, we sincerely appreciate their generous assistance.

Prologue

HOW THE MILKY WAY TURNED TO GOLD*

Once upon a time there was a cow named Bossie. She ate lush green grass and sweet yellow buttercups under a big blue sky with puffy white clouds. She gave creamy rich milk which everybody, except a few crotchety cardiologists, said was good for you.

But Bossie was not a contented cow. Bossie was a discontented cow. "Why do people keep milking me?" she said discontentedly. "All they think about is money—m-o-n-e-y—money!"

Well, Bossie had just said the magic word and—Shazam!—there stood the Dairy Fairy.

"Don't worry, Bossie," said the Dairy Fairy. "I will solve your problems. Remember my magic slogan, Everybody Needs Milk —particularly every governmental body."

So the Dairy Fairy went to the big white house where the prince lived. At the door he found a note: "Please leave three quarts of homogenized, two pints of cottage cheese, and one pot of gold."

Inside, the prince was talking to a television camera. "Hi, there," said the prince. "Golly, I guess I've always liked milk ever since my mom gave it to me with her cherry pie while dad booed the Dodgers. At the end of a hard day, I always say, 'Rosemary, bring me a shot of milk over ice,' because I believe in milk. It may sound childish but, to be perfectly candid, I also believe in honesty, integrity, and the Dairy Fairy."

Well, the Dairy Fairy was so pleased he filled the prince's order, and they had a nice long, two-hour chat about milk and how good for you it was, especially if you were running for re-election.

---

*Copyright 1974 Chronicle Publishing Company, reprinted by permission of the author.

The very next day, just by coincidence—would you believe it?—the prince raised the price of milk $300 million.

Naturally, this caused talk. In fact, some of the prince's evil enemies went so far as to whisper that he had raised the price of milk in return for the pot of gold and that there wasn't really a kindly Dairy Fairy at all!

My! What consternation this caused! An investigation was demanded by the people. A couple of dozen of the sternest judges on Capitol Hill met in solemn splendor to determine whether there was, or was not, a kindly Dairy Fairy.

But, my goodness, the clever Dairy Fairy had already visited sixteen of them to remind them how good milk was, especially if you were running for re-election, and to give them little pots of gold as momentoes of his unselfish generosity.

Well, now, the judges certainly agreed that the pots of gold had nothing whatsoever to do with the price of milk as far as they were concerned. What's more, they said, they had always believed in honesty, integrity, and the Dairy Fairy.

So everybody lived happily ever after, including the Dairy Fairy. He quit his job and became an oil lobbyist. "There's almost as much magic," he said, rubbing his hands, "in oil."

But, thanks to the wonders performed by the Dairy Fairy, Bossie was a contented cow. For instead of the people milking the cows, the cows now milked the people.

                                        Art Hoppe

Introduction

Milking the Public is a study in power, a study of how a special interest group—the dairy lobby—acquires, exercises, and maintains political and economic power in today's Washington. When relatively narrow economic or business interests seek to forge an alliance with the federal government, consumers and taxpayers are the ones who usually suffer. Business is rarely capable of consistently gouging consumers over any extended period without the assistance of government. Far more businessmen actively dislike the rigors of competition and the marketplace, therefore, than is commonly recognized. The radical historian, Gabriel Kolko, demonstrated this in his revisionist work on the Progressive Era, *The Triumph of Conservatism: A Reinterpretation of American History 1900–1916*. Kolko persuasively argues that the dominant trend in American business at the turn of the century was not toward trusts, cartels, and monopolies, but rather toward increasing economic decentralization and cut-throat competition. Big business was unable to halt these trends by itself and turned to government for relief. Kolko demolishes the cherished liberal myth that political reformers initiated the regulatory laws of the Progressive Era only to see the regulators become the captives over time of the regulated: "In virtually every case," according to Kolko, government chose solutions handpicked by business and financial interests.

The same bias against free market competition in favor of government regulation is held by many businessmen today. It is most easily observed in those industries still heavily regulated by the government. Prior to deregulation of the airline industry in 1978, for example, executives from virtually every major air carrier testified in opposition: they all *liked* regulation; they needed it to survive; only chaos would result if it were eliminated; the system required only a few minor adjustments. Not one of the airlines wanted open competition on fares or routes. Nevertheless, lacking any serious economic justification for regulation and, more important, lacking any substantial political influence in Washington, the airlines were deregulated against their will. The result—in 1978—was the most profitable year in the history of commercial aviation. Fares were reduced, passengers increased, and profits soared.

What is significant is that all these highly paid airline executives had no idea how the marketplace worked. They thought that the only way they could make profits was to keep prices high and that the only way to keep prices high was to have the government fix the prices for them. It never occurred to them that lower prices might mean more passengers and higher profits. Government regulation, not competition, was all they knew, and they fought fiercely to keep it.

The same process can be observed today among trucking executives as they oppose the push for deregulation in their industry: anarchy would result; they would be forced out of business; without government regulation, anyone who could buy a truck could haul goods anywhere there was a road at any price he could receive. Their collective hysteria is a thing to behold.

So it is with the large cooperatives who make up the dairy lobby. They like regulation; they like government subsidies; they like high tariffs on foreign dairy products; they like government-mandated prices for milk far above a free market level. They like things just as they are, and they are willing to pay to keep them that way.

Too willing. What makes this book possible is that the dairy lobby got caught trying to move too far and too fast. In another time, perhaps the fifties of Eisenhower or the sixties of Lyndon Johnson, it might not have mattered. The dairy lobby had the misfortune, however, of playing a major role in the national trauma of "Watergate." As a consequence, the dairy lobby's origins, its rise to power, and its misuse of that power were exposed for all to see.

Such an exposure provides a unique insight into how economic and political power is acquired by special interests in Washington;

how it is used and abused; and how, despite the abuses, its economic partnership with government is maintained.

The primary sources for this book are largely in the public domain—records, transcripts, and documents from the Senate Watergate Committee, which investigated abuses and illegalities in campaign contributions; from the House Judiciary Committee which, during its impeachment proceedings, investigated the role of dairy lobby campaign contributions in Nixon's dairy price support and tariff decisions; from the Watergate Special Prosecution Force, which had as many as eight attorneys working on what were termed the "milk fund" investigations; and from various legal proceedings—civil and criminal—which followed.

I. F. Stone, a journalist whose work and skeptical attitude toward government we have long admired, once complained that modern journalism was overly concerned with calling people crooks. As Stone saw it, "there have been crooks in government since the days of the Pharaohs, and there will be crooks in government as long as there is a human race and a government. . . ." Stone believed it far more important to investigate and reveal "the harm done by good men with good intentions, acting they think wisely, in the normal course of their governmental operations or their business."[1]

Stone's admonition is important because this book covers in detail the dairy lobby's "crimes"—illegal campaign contributions to the likes of Lyndon Johnson, Hubert Humphrey, Richard Nixon, Wilbur Mills, and others. Like Stone, however, we are not concerned with calling people crooks. This is not a story about a few bad apples who besmirched the dairy lobby's otherwise good reputation. We have little doubt that then, as now, the dairy lobby was composed of good men with good intentions who thought it was perfectly proper to use government power to further their private economic interests at the expense of consumers and taxpayers.

What we must remember is that the dairy lobby is more powerful today than it was before Watergate, and that it has the same purposes today that it did then. The lesson to be learned is important because it demonstrates, dramatically, the lengths to which a special interest group can safely go to influence the political process in Washington and still not endanger its economic partnership with the government aimed against consumers and the public. You can make illegal contributions to two presidents' re-election campaigns; you can make illegal contributions to other presidential campaigns including those of a vice-president and the chairman of the most powerful committee in Congress; you can blatantly attempt to influence government decisions on price supports and tariffs with campaign contributions; you can admit attempting to bribe a member of the president's cabinet; you can make illegal contributions to numerous Senate campaigns; you can make large contributions to powerful congressmen who are unopposed for re-election.

You can safely do all of these things and never once have your underlying economic partnership with government (which is all you were trying to protect in the first place) in serious jeopardy. So long as you agree to play by the rules in the future, the special alliance you have forged with the government against the public is safe.

We believe this is a story worth telling—and remembering.

Part 1

Acquiring Power: Milk, Money, and Monopoly

We are just trying to do what business and labor have been doing for 100 years. If you're asking me if we were buying favors, the answer is no. We are not in the business of buying people. We are just trying to promote our product with the help of our friends. This is a cruel world. You've been around long enough to know you don't get nothing you don't pay for.

W. R. "Preach" Griffith, Chairman
Trust for Agricultural Political Education

1

Of Milk and Money

The dairy lobby's story covers ten years and four presidents, and moves from high drama to low comedy, and from paying seamy subsidies to one president to playing a major role in the impeachment of another. From literally out of nowhere, the dairy lobby during the years 1967 to 1977 created for itself a major position of prominence and power at both ends of Pennsylvania Avenue.

In climbing to the top of the greasy pole, many lobbyists, lawyers, and politicians alike would be investigated, indicted, and convicted. Some would go to jail. Yet when the smoke cleared, the dairy lobby was still sitting on the top of the pole, still doing business as usual. Congress was more firmly in its grasp than ever. Its relations were well cemented with a new president who had himself become wealthy through federal farm subsidies.

The dairy lobby's rise to national political power during the late 1960s and early 1970s was inextricably involved with "Watergate." The attitude and approach of the dairy lobby proved more than a match for politicians, even in the moral atmosphere of those times. Dairy lobbyists actually believed they could buy presidential decisions with campaign contributions, legal and otherwise. They believed the only way they could persuade Congress to listen to dairy farmers was with the "jingle of hard currencies." They believed they could bribe the secretary of the treasury. They believed campaign contributions could buy their way out of major government antitrust litigation. Newly arrived to their position of major economic and political power, they believed that this was the way our government worked. The sad, even scary, part is that they were right more often than not.

Milk in this country is big business. Retail dairy sales annually exceed $17 billion. Over 50 percent of this figure is directly attributable to the cost of raw milk.[1] To understand why the dairy lobby exists, it helps to know that the production of Grade-A raw milk (the kind which comes directly from cows) in this country is controlled by a number of regional milk cartels which are formed and administered by the Department of Agriculture. There is no free market price for raw milk produced by dairy farmers belonging to these cartels. The USDA fixes a minimum price—far above what a free market price would be—which milk processors must pay to farmers operating within the cartel areas. Milk-price supports—the announced price at which the Department of Agriculture will purchase all unsold (surplus) milk—play a major role in keeping cartel prices artificially high.

By 1967, 86 percent of all dairy farmers in the USDA-sponsored milk cartels had become members of milk producer cooperatives. This is a significant figure because from 1967 to 1977, over 170 local milk producer cooperatives with a combined membership of 70,000 dairy farmers producing over 26 billion pounds of milk annually merged into four large multimarket regional cooperatives located in the central United States. These four regional "super cooperatives" attained a monopoly-level control of more than 75 percent of the raw milk supplied to such major fluid-milk markets as Chicago, Madison, Cleveland, Toledo, Cincinnati, Indianapolis, Louisville, Houston, Dallas, San Antonio, Memphis, Oklahoma City, Omaha, and Minneapolis–St. Paul.[2]

This strange alliance of private monopoly power firmly grafted onto government-sponsored cartels has produced a milk-marketing system which benefits the few at the expense of the many.

For consumers and taxpayers, it is a bad system and it is wrong. It spends millions of dollars on supports to keep milk prices high when those same high prices force demand down to a point where thousands of dairy farmers have to go out of business every year. It fixes milk prices that result in $400 million in consumer overcharges each year and then allows private monopolies to charge even higher prices. It is a system, initially designed for the relief of small dairy farmers, which disproportionately confers its benefits on large wealthy dairy farmers who need them the least.

The purpose of the dairy lobby is to keep things just the way they are. As one prosperous dairy farmer recently conceded to *Business Week,* he and his cohorts are better off than most other farmers in the country, and they intend to "remain satisfied."

The dairy lobby consists, chiefly, of the political action committees of the nation's three largest milk producer cooperatives to which dairy farmer members of those cooperatives contribute. As John Connally explained it to Richard Nixon in 1971 in the privacy of the Oval Office: "These dairymen are organized; they're adamant, they're militant. They, uh, very frankly, they tap these fellows—I believe it's one-third of one percent of their total sales or $99 a year whichever. . . .Oh, it's a checkoff. No question about it. . . .And they, they're massing an enormous amount of money that they're going to put into political activities, very frankly."[3]

Frank Wright was the Washington correspondent for the *Minneapolis Tribune* during the Watergate era. His investigative reporting of the dairy lobby's massive campaign contributions to the president's re-election was a constant irritant to the Nixon Administration. Now the *Tribune*'s managing editor, he summarized in a letter to the authors in 1978 the dairy lobby's operations in those days. It was, says Wright, "a classic case of special interest trying to buy its way to power. What made it different from so many others was that the dairy guys were so new at it, so naive and so gross. They really were right up front about it; you sell, we buy. . . . Every time I did a story in those early months, detailing new payments, they'd ask me for additional copies of the story. They were so proud of what they'd done and so happy about the publicity that they wanted to send copies to all their regional and local board members to demonstrate how effective their new political pay-off scheme was. I couldn't believe it. Needless to say, they smartened up when it dawned on them that not everybody thought what they were doing was such a great thing."[4]

Since the dairy lobby smartened up, it would be a mistake to think that it is a Watergate phenomenon whose power vanished with Nixon's resignation. If anything, the dairy lobby is more powerful today. For example, while Gerald Ford's administration was openly hostile to the dairy lobby, Jimmy Carter has given it a warm embrace. Less than three months into office, Carter, like Richard Nixon before him, ignored the law and made a federal milk-price-support decision based *solely* on what Nixon liked to call "traditional political considerations"—i.e., paying off a political campaign promise made to dairy interests.

The dairy lobby owes its increased political strength today to organization and money, especially money. You do not find the dairy lobby driving tractors or herding animals through the streets of Washington as did their fellow farmers during the winters of 1978 and 1979. When dairy farmers want another subsidy from the federal government, they go to Capitol Hill quietly and smoothly with well-paid lobbyists and lawyers who carry brief-cases full of money. Next to the American Medical Association, the dairy lobby gave more campaign contributions to House and Senate candidates in 1976—over $1.3 million—than any other interest group. They gave more than the AFL-CIO, the United Auto Workers, the maritime unions, the oil and gas lobby, or the education lobby.

There has been endless debate about whether special interest groups are buying influence and access to legislators with their campaign contributions (as their detractors assert) or are merely supporting candidates who are ideologically attuned to positions favored by the interest group. Compare organized labor with the dairy lobby. In 1976, labor's ratio of contributions to incumbent congressmen over challengers was only 2.9 to 1, the lowest of any of the major special interests. The dairy lobby's ratio, on the other hand, was 7.3 to 1, far and away the highest of all special interest groups. As evidenced by its support of challengers, a good case can be made that labor spends its money ideologically—to those candidates, challengers or incumbents, who share labor's views. Why else spend so much money on challengers who traditionally lose 9 times out of 10 anyway? Not so the dairy lobby. It wastes no time with losers. Its money goes straight to incumbents and that demonstrates its intent—the dairy lobby blatantly buys access and votes on the relatively few and narrow economic issues with which it is vitally concerned.

As Richard Nixon observed in a meeting with milk producer cooperatives: "Uh, I know . . . that, uh, you are a group that are politically very conscious. Not in any partisan sense, but that you realize that what happens in Washington, not only affecting your business, but, affecting the economy, or foreign policy and the rest, affects you. *And you're willing to do something about it.* And, I must say a lot of businessmen and others that I get around this table, they'll yammer and talk a lot but they don't do anything about it. *And you do, and I appreciate that. And, I don't have to spell it out"* [emphasis added].[5]

Nixon was right. What happens in Washington does affect dairy farmers. It has been affecting them in a direct bottom-line fashion for almost fifty years.

To appreciate and understand the dairy lobby's rapid rise to political power during the late 1960s, its abuse of that power during the early 1970s, and its consolidation of that power after the smoke from Watergate had cleared, a short political history of milk is necessary. For, surprising as it may seem, the milk industry has almost always been a subject of political controversy and government regulation.

2

The Golden Age and the New Deal

It is a myth of the dairy lobby that the economic survival of dairy farmers depends upon government subsidies and government price-fixing, or, for that matter, upon illegal pay-offs and political campaign contributions. Leland Totman and his father are proof that this is not so. They are dairy farmers from New England, a place not nearly as profitable for dairy farming as Wisconsin, Minnesota, or upper New York State. Yet Lee Totman is an excellent farmer, once named Massachusetts Outstanding Dairyman of the Year. After taking over the farm from his father, he raised the annual milking average of each cow by over six thousand pounds. His cows now produce over nineteen thousand pounds of milk per year, more than twice the national average. The Totmans' success was explained by a neighbor: "Farmers basically compete with one another to do things the most productive way, and the smart farmer is the one who adopts the coming methods five years before his neighbors do. Lee is like that. His dad was too. They're farmers' farmers."[1]

Totman welcomes the technological changes in agriculture which have reduced the U.S. farm population to less than 5 percent of the total. "To me that's business," says Lee, "free enterprise—all that, isn't it? The ones who go out must think it's better to go do something else—go on welfare maybe, or else the state makes so many regulations they are forced out. And another thing. You think everybody who doesn't farm ends up on welfare or working in a factory? Look at my brother, Conrad. He's a college professor. If there were no milk machines or tractors or artificial insemination or automatic gutter scrapers, you know what Conrad would be doing? He'd be in the barn with a shovel."[2]

In his own way, Lee Totman has identified one of the central facts in agricultural politics in this century—the inevitable redistribution of manpower from agriculture to other occupations caused by the increasing industrialization of American society. The agricultural population in this country has been in a steady decline since the late nineteenth century, and that decline continues unabated today. This fact is essential to making sense out of governmental policies in agriculture because organized farm interests have long used this decline as one of their major arguments in persuading politicians to pass subsidies for farmers. Look at all the farmers going out of business, they argue; if government does not do something to help them, no one will be left to produce the food.

Nobel Prize winning economist F. A. Hayek has observed that the popular support which the resulting government farm subsidies receive is occasioned by the erroneous belief that *all* of agriculture, rather than its less productive members, is unable to earn an adequate income. Hayek has written that the increase in agricultural productivity, combined with the general inelastic demand for food in an advanced industrial civilization, means that if farmers are to maintain their average income, marginal farmers must leave the land. Nevertheless, so long as the readjustment in the agricultural population takes place, there is no reason those productive farmers remaining in agriculture should not continue to benefit as much economically as the rest of society. Hayek suggests that the natural reluctance of marginal farmers to shift to other occupations causes the market price of agricultural products to fall "much lower before the necessary readjustments [are] effected then they would have to do permanently."[3]

Marginal farmers will not be induced to leave farming unless their agricultural incomes are reduced relative to what they could earn in town. If enough marginal farmers hang on and continue to produce, the greater farm production which results will naturally lead to overall lower prices. Yet the lower prices will *never* be low enough to drive efficient, successful farmers like Leland Totman and his father off the farm.

Nevertheless, it is this myth that *all* farmers are unable to earn adequate incomes which has served as the core of government agricultural policies in this century, whether government price supports for commodities or federal regulation of milk.

Yet, all the government subsidies have not been able to keep marginal farmers in business. Even today, people are continuing to leave farming just as they have throughout this century. In the meantime, the government subsidies of various types to agriculture continue, and one wonders who is benefiting from them. Certainly not the consumers or taxpayers, who are *paying* for it, both in higher taxes *and* higher food prices. Certainly not marginal farmers who continue to go out of business.

The answer is that only successful farmers benefit from the government subsidies. The ones who need the subsidies the least benefit the most. This is because subsidies are given in a non-discriminatory fashion to farmers rich and poor alike and are based on their total production. Successful farmers invariably have larger farms, more crops, bigger herds than marginal farmers. Their share of government subsidies is proportionately far greater than the relative pittance received by marginal farmers who are, because of their small and inefficient size, barely managing to stay in business. .

As for the politics of all this, the secret is to make Congress believe that *all* farmers are marginal and in imminent danger of going out of business. As we shall see, it is not that hard to do.

A good frame of reference for examining the origins of the dairy lobby's immense political power today is American agriculture between the Civil War and World War I. This was a period of immense growth in the American economy, and agriculture shared in that growth: the number of farms tripled; the number of acres of farmland doubled; and net farm income increased more than four-fold. Despite this growth, two stark figures stand out. Farm population *decreased* from 60 percent to 35 percent of the national total, and farm income *decreased* from 31 percent to 22 percent of national income. In other words, the absolute increase in agriculture did not keep pace with the increase in the rest of the economy—it fell behind.[4]

The reasons for the growth in agriculture are many—developments such as the McCormick reaper, the Marsh harvester, the use of steam power in wheat threshing, and the disc type rotary harrow. Other factors included the vast increase in scientific farming, and the accompanying dissemination of scientific knowledge of efficient and advanced farming methods through public and private agencies. The result was that only farmers who could adapt and become efficient managers prospered because they had a distinct advantage over farmers who stuck to the old ways.

On a political and economic level, there was much farm discontent during this era despite increased income, greater output, and new technology—chiefly because the farmers' share of income as a percentage of the national total had decreased. Nevertheless, per capita farming income increased during this period from $70 to $112 a year, and likewise per capita wealth increased from $416 to $755.[5] Farmers were particularly hurt by the deflation which followed the Civil War—by 1896, farm prices had fallen to approximately 25 percent of the 1865 level, while nonfarm prices dropped only to one-third of their Civil War level.

In response to this deflation and their decreased share of national income, the farmers found many villains: railroads, grain-elevator operators, farm implement manufacturers, and banks. Naturally enough, farmers demanded that the Government take action to stop the deflation.[6] The farm discontent during 1865-1913 gave rise successively to the Grange movement, the Farmers Alliance, and the Populist Movement. This is politically relevant today because the real villain was an economic problem, the same economic problem which has continued to plague American agriculture throughout this century—overproduction.

Between the Civil War and World War I, the tremendous expansion of agriculture resulted in production far exceeding the total demand for farm products. Consider wheat—per capita wheat output increased from 39.22 bushels in 1859 to 53.79 bushels in 1879, and 55.89 bushels in 1889. The resulting drop in the price of wheat was drastic—from $2.06 per bushel in 1866 to $.49 per bushel in 1894.[7]

One economic historian has observed there were six possible solutions to the problem of agricultural overproduction: increase demand; decrease supply; subsidize farmers; allow market forces to eliminate inefficient farmers; decrease production costs; raise prices through inflation of the money supply.[8]

Because farmers are small economic units and are price-takers not price-setters, they had no power individually to increase overall demand for, or decrease overall supply of, their products. The third alternative, subsidization (one of the solutions subsequently adopted in the twentieth century), was politically unpalatable in the superficial laissez faire era of the late nineteenth century. However, allowing market forces to eliminate inefficient farmers was as politically unpopular then as it is now.

By a process of elimination, therefore, agriculture's political and economic goals during this period were reducing production costs (e.g., railway gates and grain elevator charges) and inflating the money supply (e.g., the Greenback and Free Silver Movements). The farmers' complaints are instructive and echo current farm attitudes toward government: "The imposition of high charges by corporate organs of wealth . . . stories of bribed legislators . . . purchased domination of the agencies of government."[9] Some of the criticisms were not without merit. Localized monopoly situations did occur with railroad rates and grain elevator charges. The general deflationary era did benefit the banks. Even without localized monopolies, price-fixing among railroads or grain elevators was quite common. Yet with all this, overproduction by the farmers is inescapably the primary reason for their overall economic plight.

The dairy industry as a commercial entity—something apart from milk production and consumption solely on the farm— started in the East where the growth of cities had created a market for fluid milk and butter. By the time of the Civil War, the industry had spread west to northeastern Ohio, where a thriving cheese industry had been established. After the Civil War, the commercial dairy industry expanded into the Dakotas and Minnesota, where the weather had proved so inhospitable to crops that dairy farming became a means of coping with the unpredictable harshness of the climate.

As a practical matter, fresh milk had to be marketed locally. Before pasturization and refrigerated shipping, fresh milk had a very short life and quickly spoiled. This resulted in small and isolated markets and a dairy farmer's reputation for quality was therefore most important.

Technology played a large role in the infant industry. The nation's first mechanical refrigeration plant opened in Boston in 1881. Refrigerated shipping from Bangor to Boston was started that same year by the Maine Central Railroad. The increasing use of refrigeration combined with faster trains enabled milk to travel greater distances. Massachusetts farmers shipped 61 million quarts in 1870 and doubled that figure ten years later. By 1895 Boston was receiving milk by train from areas as far away as Vermont, New Hampshire, and northeast Connecticut. In addition, some Connecticut milk was going to New York City.[10]

Dairy farmer cooperatives in this period were quite primitive. Jesse Williams of Oneida County, New York was an early venturer into cooperative marketing of milk products. Williams was an experienced maker of quality cheese at a time when most American cheese was not very good. In the spring of 1851, one of Jesse's sons had married and gone into farming on his own. Jesse had contracted for the two of them to sell cheese at seven cents a pound, more than other dairies in the area were paying. But the son was worried that he could not produce acceptable cheese. So Jesse and his son decided that the milk from the son's farm would be delivered daily to Jesse's milk house, where he would supervise the cheese-making. This proved to be such a good idea that other farmers also started shipping their milk to Jesse, who would supervise the cheese-making, sell the cheese, and share the profits pro rata with all the farmers shipping to him.[11]

An early cooperative venture for marketing raw milk to dairies was the Old Colony Milk Producer Association organized in 1877 in Plymouth, Massachusetts. By 1891, another milk producers association in Massachusetts was claiming credit for gaining a one cent increase in the price of raw milk. At the turn of the century, the Springfield Co-operative Milk Association was marketing approximately two thousand gallons of milk a day.[12] In the milk-producing area serving New York City, the Five States Milk Producers Association was organized in 1898 and attempted, with little success, to raise the price farmers received for their milk.

The early experience of such cooperatives was hardly encouraging but established the foundation for future growth and success. The first milk strike by a cooperative to enforce its pricing position started in Chicago. It was successful and spread to southern Illinois, where it failed because there were enough dairy farmers not under the control of the cooperative who were willing to supply milk to the dairies at prices less than those of the cooperative. The concept of milk strikes spread to New York, Boston, and Cincinnati but failed to achieve any lasting success:

> The limited successes of the early cooperative associations can be attributed to many factors, including imperfect flow of market information and absence of cooperation between essentially isolated markets. In the early 1900's, there was no centralized collection or publication of data about milk prices and production figures. There was not even potential for easy communication between farmers or cooperatives: the phone and radio were new ideas at the time. Markets were further isolated because milk could not travel long distances. Thus, there was both physical and economic isolation of milk markets.[13]

Another problem faced by the early cooperatives was laws prohibiting them from joining together in cooperative marketing associations. Many states and the federal government believed such associations to be illegal conspiracies in restraint of trade. The experience of the Chicago Milk Shippers Association is typical of how that belief worked in practice. It was formed in 1891 and had approximately fifteen hundred members. The association received milk from its members, guaranteed them payment for milk it sold, and (importantly as it turned out) set the price for which it would sell their milk. One of its customers was a dairy owned by Charles Ford, who managed to run up a $438 bill. When Ford failed to pay, the association sued him. Ford never denied receiving the milk but nevertheless resisted payment. The case was eventually heard by the Illinois Supreme Court, which ruled that there was a conspiracy to restrain trade and fix prices between the association and its dairy farmer members, and that hence the contract between Ford and the association was void.[14] Ford was happy since he got the milk but did not have to pay for it. The association was not happy, and neither were its members.

The first two decades of this century were the most prosperous ever for American farmers—and the American public has been paying for it ever since. From 1900 to 1910, the index of wholesale prices for all farm products increased almost 50 percent, considerably more than other segments of the economy.

A combination of several factors caused this upsurge in farmers' prosperity. By 1890, the frontier was closed and all good agricultural land had been settled. Further, the farm population had reached its peak and would increase by only 4 percent from 1900 to 1910. The nonrural population as a whole increased ten times that amount. Hence, there was an expanding market for farm products, demand having increased much more rapidly than supply.[15]

World War I in its early years sustained the growth achieved between 1900 and 1910, and the United States' entry in the war caused a further increase in farm prosperity which resulted in a rise in net farming income from $4.39 billion in 1915 to $9.8 billion by 1919. One economic historian observed that "it would be difficult to find an example in history of so large a class of people rising so rapidly from relative poverty to comparative affluence as did large sections of American farmers in the twenty-five years after 1896."[16]

The prosperous years from 1900 to 1920 abruptly ceased after World War I, and American farmers turned to the government to insulate and protect them against the free market taking away what it had granted them during the earlier "golden" period. The end of the war literally brought economic reality back to the farms. The increased demand, production, and prices characteristic of the war were replaced by sharply declining farm prices and increased prices for farm supplies.

The sudden drop in demand after the war had several causes. With war's end, the United States government stopped financing European purchases of American food, even though American agriculture had over-expanded to satisfy the artificial war demand. Rival food suppliers from Australia and Argentina were once more able to secure shipping for their products to Europe, and Europe's own agricultural production was reviving. At the same time, American consumers were losing purchasing power in the 1921 depression. The index of prices received by farmers for all products fell from 215 in 1919 to 124 in 1921. Supply prices, however, fell only from 202 to 152. The result was that during the 1921 depression, farmers' purchasing power was only three-quarters of what it was before the war.[17]

Even though the farm population was declining, gross income from farming was increasing and the average farmer was receiving a larger return, the proportion of the national income going to agriculture in the 1920s decreased just as it had *in every decade since the Civil War*. Faced with this proportional decline, farmers turned to the government and claimed this inevitable trend of an agricultural society in the midst of transformation to an industrial one was the reason the federal government should step in and protect farmers from competition.

During the 1920s, organized agriculture had two major political goals. One goal was a two-price agricultural system—an artificially high price for the American market, protected by an absolute tariff, combined with an essentially free market price for surplus distributed abroad. The second goal was antitrust immunity for individual farmers and farm groups, so that they could join together and effectively compete with suppliers and middlemen.

The price-fixing scheme—later known as the McNary-Haugen Bill—was born in a moment of financial desperation when the Moline Plow Company of Moline, Illinois, was thrown into bankruptcy. Its managers, George N. Peek and Hugh S. Johnson, correctly calculated that the farm machinery business was not going to prosper unless agriculture prospered. Both Peek and Johnson had served on the War Industry Board and preferred the safety of government subsidy to the rigors of competition. Accordingly, in late 1921, Messrs. Peek and Johnson presented a plan to the American Farm Bureau Federation to create prosperity on the farm through government action (at the expense of the rest of the economy). The plan proposed a governmental framework for selling farm products for domestic consumption at a "fair exchange value" based upon the ratio between industry and agriculture in the golden period of 1910 to 1914, and for selling the surplus on the world market for whatever it would bring. The details of the Peek-Johnson plan provided for achieving a government guarantee of farm prices through tariffs which would protect an artificially high price at home while the resulting surplus was exported at a market price.[18]

The biggest supporter of the Peek-Johnson plan in the Harding Administration was Secretary of Agriculture Henry C. Wallace, Jr. For many years, Wallace had been the editor of the farm journal *Wallace's Farmer* with an editorial policy consistently favoring bigger and better agricultural schools and government regulation of the railroads and opposing the War Food Administration and its administrator, Herbert Hoover.

Wallace's animosity toward Hoover carried over to the Harding Administration in which Hoover served as the secretary of commerce. He was considered no friend of the farmer. In 1920, when Hoover's name was being bandied about in prepresidential maneuvers, he was called by a Grange representative "the most objectionable to the farmers of this country." Wallace himself had said of Hoover in connection with his work in the War Food Administration that "his dealings with hogs and milk and beef producers gave evidence of a mental bias which causes farmers to thoroughly distrust him. They look upon him as a typical autocratic big businessman."[19]

As secretary of agriculture, Wallace was continually engaged in bureaucratic infighting with Hoover. Hoover apparently believed that the Department of Agriculture should be limited to telling the farmer what crops to grow, while Hoover's own Commerce Department told the farmer how to dispose of the crops. Wallace, on the other hand, clearly favored the 1920s trend of farmers' seeking relief from their economic position by political means, rather than allowing prices to be determined by the play of natural consumer preferences: "Unless farmers as a class get busy and *fight* for their rights, we in the department will not long be able to take a national point of view because the point of view of other interests [consumers] would dominate us."[20]

Wallace prepared a draft bill based upon the Peek-Johnson plan and had it introduced in January 1924 by Senator Charles McNary of Oregon and Representative Gilbert Haugen of Iowa. Hoover strongly opposed the McNary-Haugen Bill, which was defeated in the House in 1924. In October of 1924, after the defeat of McNary-Haugen in the House, Secretary Wallace died and was replaced by W. M. Jardine, an ally of Hoover.

However, McNary-Haugen did not die with Secretary Wallace. Secretary Wallace's son, Henry A. Wallace, called by Arthur Schlesinger, Jr. a "brilliant experimental geneticist and a talented farm economist"[21] in his own right, took up the battle from his post as editor of *Wallace's Farmer*. Unlike his father, he had long advised farmers that the only way to maintain income in the free market was to reduce the size of their crops. While he was critical of those who advocated the substitution of price-fixing legislation for the free market, he nevertheless continued to support McNary-Haugen (despite its implicit price-fixing nature) and was instrumental in developing a western/southern alliance, which resulted in its passage by Congress in 1927, only to be vetoed by President Coolidge. McNary-Haugen was passed again in 1928 and once again vetoed by Coolidge. Farmers were prospering generally during the 1920s though less than the rest of the country, however, and their lobbyists and supporters in Congress were never able to muster enough votes to override the two vetoes.

The second major legislative effort by farmers in the 1920s was passage of the Capper-Volstead Act, exempting the formation of farmers organizations from antitrust laws. Even though many farmers and small dairies had their own retail milk customers, the Boston and New York milk markets had, by 1910, become dominated by a few large dairies which were able to control both the retail price of milk to consumers and the wholesale price for raw milk to be paid to farmers. If the large dairies met resistance from the farmers, they simply extended the geographic area from which they purchased raw milk. In Boston, as we have seen, this led to the formation of the Boston Cooperative Milk Producers Union. In response to pressure tactics of the large dairies, the ᴮoston Union waged the so-called Boston Milk War of 1910 with weapons both legal (milk withholding) and not so legal (intimidation and milk dumping). The dairies' response was to go straight to court, and in 1911 the Milk Producers Union was declared an illegal combination in restraint of trade.[22]

The Capper-Volstead Act is generally considered the Magna Carta for modern agricultural cooperatives. Since agricultural cooperatives form the backbone of the dairy lobby whose corrupt influence on the government will be examined in subsequent chapters, it is important to examine Congress's intent in authorizing an exemption from antitrust laws for agricultural cooperatives.

Congress's main concern in the debate on the Capper-Volstead Act was to allow farmers to form cooperative associations which would essentially give them the advantages of a corporate structure free from the threat of antitrust prosecution, such as had occurred on the state level after the Boston Milk War. The purpose of the bill was summarized by its co-author, Representative Volstead:

> Business men can combine by putting their money into corporations, but it is impractical for farmers to combine their farms into similar corporate form. The object of this bill is to modify the laws under which business organizations are now formed, so that farmers may take advantage of the form of organization that is used by business concerns.[23]

Ironically, in light of what was to occur almost fifty years later, the possibility was considered that milk might eventually be controlled by monopolistic organizations. As one senator, otherwise a strong supporter of the cooperative movement, observed:

> [M]onopolies can be organized with reference to certain limited products of the farm that are grown only within a very restricted area or that are perishable in character and will not stand shipment from one end of the country to the other. I may say here . . . that milk is one of those things the supply of which to the great industrial centers, to the great populous cities of the country, can be monopolized . . . just as easily as not under this act.[24]

Today, it is commonly recognized that Congress, in enacting the Capper-Volstead Act, was reacting to several political-economic perceptions. First, it believed there was a tremendous imbalance in power between farmers and the corporations they had to deal with—both dairies and suppliers of farm goods—thus placing the farmers at a marked disadvantage in selling their products. Second, Congress believed that authorizing the creation of cooperative associations was the answer to this imbalance of power, the correction of which was supposed to solve the farmers' income problems.

Capper-Volstead, however, must be kept in perspective. It was a limited response to a limited problem. It simply granted the farmers the right to organize into cooperative associations without being subject to the antitrust laws. It was not intended to be a license to violate the antitrust laws with impunity, to monopolize at will. Congress intended only to give to farmers the same advantages of collective action and benefits of size enjoyed by investors in corporations.

The concept of classified pricing is the central element in understanding today's milk cartels. Classified pricing systems established by early cooperative efforts at bargaining gradually prevailed in most major markets and extended through the 1920s and early 1930s. Classified pricing has two separate aspects: (1) the price charged by the cooperative to the milk processor for raw milk and, (2) the amount returned by the cooperative to the farmer member.

First, the co-op price to the milk processor is based on raw milk's use *after* processing, the price for raw milk processed into fluid milk being higher than that for milk processed into manufactured milk products like cheese or milk powder.

Second, a cooperative pays each of its farmer members the same price for his raw milk, regardless of whether it ends up as fluid milk or cheese. This is typically referred to as a "blend" price and is a weighted average of the various prices received by the cooperative for its farmer members' raw milk, depending upon its use.

This dichotomy in pricing of a fungible commodity like milk will generally not occur in a free market, except when there is a shortage of Grade-A raw milk (Grade-A, subject to stricter sanitary standards and hence more costly, is the only raw milk acceptable to be processed into fluid milk. Cheese and milk powder can come from either Grade-A or Grade-B raw milk). Because of its perishable nature, fluid milk cannot be stored for long periods. The demand for fluid milk is fairly constant and relatively inelastic —a change in price does not significantly affect demand. During a shortage, a fluid-milk processor will pay more for raw milk than he otherwise would during normal conditions because he must supply his customers and knows he can pass on the price increase. A manufacturing milk processor, on the other hand, faces an elastic demand for his products—changes in price will affect demand. Further, most of his products, such as cheese or powdered milk, can be made from Grade-A or Grade-B milk and can be safely stored for relatively long periods. During a shortage, he has no incentive to pay higher prices for Grade-A raw milk because he cannot pass on the increase to the consumer and, unlike a fluid-milk processor, he can afford to wait for Grade-A prices to fall because his product will not perish in the interim.

Dairy cows fail to appreciate these market nuances. They produce more milk in the spring (the "flush" season) than they do in the late fall and early winter (the "short" season). Accordingly, raw milk ordinarily commands a higher price from fluid-milk processors in the short season than in the flush season, when it tends to bring the same price from fluid milk processors as it does from manufactured-milk processors because of the natural surplus condition.

Prior to the 1930s and the advent of federal milk regulation, this natural production cycle led to intense competition among dairy farmers to find a fluid-milk processor to ship to on a year-round basis, because the farmers would automatically receive higher prices during the short season. The more intense the competition among dairy farmers to find a fluid market, however, the lower the "short" season high prices would be, and, hence, the lower the overall price for raw milk on a year-round basis.

The purpose of "classified pricing," as introduced by the early cooperatives, was quite simply to eliminate competition among dairy farmers for fluid milk outlets and to impose a year-round high price to be paid by fluid milk processors. By joining a cooperative association, which would exclusively market milk for them, dairy farmers would no longer compete for a fluid market for their milk. Whether the cooperative sold its members' milk to a fluid-milk processing plant or to a manufactured-milk processing plant, each member would receive the same average price for his milk as any other cooperative member.

However, classified pricing had enforcement problems. Fluid-milk processors attempted to subvert the system by finding cheese plants which would purchase raw milk at a lower price ostensibly for their own use and then resell it to the dairy for fluid use at a price less than the cooperative was charging. Likewise, any independent farmer who dealt directly with a fluid-milk processor in a classified-pricing market could usually receive more for his milk than the blend or average price paid to cooperative members. In this way, the processor could often pay an independent farmer more than the blend price for his raw milk and still pay less for its raw milk than a processor buying from a cooperative. Yet the price received by that processor's independent farmers could be greater than the blend or average price paid to members of the cooperative, since a cooperative, because of over-production, often found itself with a large surplus, which it disposed of at the manufacturing rate to cheese plants, thus lowering the average price paid to its members.

Milk processors were able to find such independent farmers who were willing to sell their milk at manufactured-milk prices, thus undercutting the cooperative's higher price. They were able to do this despite the willingness of milk-producer cooperatives to employ milk strikes and violence. Accordingly, throughout the 1920s and the early years of the Depression, the two-price system was unstable. Cooperatives which supplied milk to large urban markets regarded the frequent breakdowns in the two-price system as socially undesirable and chaotic. Their disingenuous argument was that milk strikes and their accompanying violence (caused by the efforts of cooperatives and their members to enforce the discriminatory two-price system) were an undesirable consequence of competition. Their solution—a not atypical reaction of businessmen confronted with competition and a genuine free market—would be to have the federal government suppress the competition.[25]

During the 1920s, both milk processors and farmers attempted to keep the price of fluid milk up, but this naturally resulted in a decrease in the amount sold. As the price of milk used for cheese and manufacturing purposes fell, however, farmers who regularly sold Grade-A milk to manufacturing plants undercut the price of fluid milk in major markets. In Chicago, for example, a strong milk cooperative, the Pure Milk Association (PMA), and a close-knit group of major milk processors made unsuccessful efforts to fix the price of fluid milk. Farmers who were not members of the PMA undercut the fixed price in several ways. First, they sold milk directly to the public at a lower price than that charged by the major milk processors. They were able to do this despite pasteurization requirements because of the introduction of small-scale pasteurization equipment in the late 1920s. Second, smaller milk processors offered to buy raw milk from nonmember farmers at a price higher than that for manufactured milk and lower than that being offered by the PMA. These small milk processors then attempted to increase their market share by selling fluid milk to consumers below the price of the large milk processors.[26]

As early as 1928, some milk cooperatives had attempted to control all the production of milk in their markets so as to keep prices high. These efforts failed because the co-ops were unable to secure the support of all farmers in a given market. As long as some farmers remained outside of the cooperative, any effort to reduce or control production would be frustrated because nonmembers would increase their production anytime the cooperative was able to raise the price of raw milk artificially. The cooperatives recognized their inherent weakness operating under free market conditions and made various attempts, from 1929 through 1932, to secure government assistance to suppress competition. It is important to keep in mind that even during the prosperous 1920s, marginal dairy farmers were still being forced out of business, as the agricultural population continued its inevitable decline vis-à-vis the nonfarm sector. During the boom years of the 1920s, the displaced dairy farmers usually had no problem finding work off the farm. Legislative attempts to enforce the cooperative efforts to suppress competition were, therefore, unsuccessful.

The onset of the depression changed this, and the consequences were felt by all dairy farmers. Instead of leaving their farms, marginal farmers stayed and tried to eke out a subsistence because there were no jobs available in the cities. Price competition by marginal dairy farmers became intense. The classified-pricing systems of the cooperatives could not withstand the competition, and they collapsed.

The competitive situation had become so critical to some dairy farmer cooperatives that they commenced a series of milk strikes to win higher prices for fluid milk. The strikes were self-defeating; they only caused more milk processors to go farther afield and lure raw milk away from cheese plants in rural areas. When this happened, the striking farmers retaliated with violence. In Chicago, where the Pure Milk Association was conducting such a strike, rural manufacturing plants shipping milk into the Chicago market in defiance of the milk strike were bombed. Similar violence occurred elsewhere in the Midwest.

Sioux City, Iowa, is an example. On August 11, 1932, the Sioux City Milk Producers Association ordered a milk strike and declared an embargo on the entry of nonmembers' milk into the city. Pickets were established on all ten highways into Sioux City. Trucks carrying nonmembers' milk were flagged down and turned around. The mood of the picketers soon turned ugly, when blockade runners made their appearance and nonmember milk began arriving in Sioux City. The groups surrounding Sioux City increased in size up to one hundred men. Physical obstacles were placed across the highways, milk trucks were stopped, milk cans were ripped open, and the raw milk was dumped in ditches. Law enforcement authorities were virtually helpless. Sheriff's deputies were unable to remove the co-op barricades, which by that time had become quite effective. The pickets had developed a system of railroad ties and spiked planks with ropes attached and, when a milk truck approached, they would be drawn across the highway. Some trucks that made it through the barricades were chased by farmers into the city until the farmers were driven off by police. On one occasion, freight trains carrying milk were halted.[27]

Council Bluffs, Iowa provided an even uglier episode. On one occasion, a group of fifty sheriff's deputies used tear gas to clear out a group of pickets. On another occasion, deputies arrested a dozen farmers who refused to break up a picket camp. By the night of April 25, 1932, Council Bluffs' jail contained forty-three strikers, and the stage for confrontation was set. From all over Iowa, truck caravans of dairy farmers moved towards Council Bluffs. The sheriff prepared for the siege and called in support from Omaha police. Machine guns were mounted. By morning, three thousand farmers surrounded the county jail threatening an attack to free the arrested strikers. Faced with the prospect of being able to halt such an attack only with his machine guns, the sheriff relented and released the strikers on bail when one of the farmers offered to pledge his unencumbered property as a bond.[28]

Violence continued to characterize the dairy co-ops' actions and the authorities' response. Near Sioux City, eleven deputies were hurt in a brawl with farmers. On August 31, near Cherokee, Iowa, a group of farm pickets were attacked by unknown assailants with shotguns who wounded fourteen and killed one. The county sheriff and an officer of the local bank were tried but not convicted of the attack.[29]

By May of 1933, dairy farmers generally were not in much better shape than the rest of the country. In 1929, prices for dairy products were 103 percent of what they had been during the golden era of 1910 to 1914. By May of 1933, dairy prices had dropped to 61.7 percent of their level during 1910 to 1914. Moreover, the problems of the dairy farmers went beyond price. Like other farmers, they were faced with the effects of bank failures and mortgage foreclosures.

Accordingly, when physical violence and milk strikes failed (essentially at the point of a gun) to coerce more money from consumers, dairy cooperatives turned to Congress to see if the government might not be successful in forceably transferring money from consumers to farmers.

The National Milk Producers Federation, a lobbying group consisting of most of the country's dairy cooperatives, told the Senate Agriculture Committee in 1933 that milk processors sold milk too cheaply to consumers, implying that consumers would have bought just as much milk, even if the price had not been as low:

> The price of milk in the cities has fallen tremendously [and] that fall was not so much due to the lack of consumer buying power in the cities for our milk as it was due to the lack of coordination among the [milk processors] themselves, the result

being a series of disastrous price-cutting tactics in many cities of this country to far below what there was any consumer demand for it to go down to.[30]

In other words, those milk processors should have been price-fixing and gouging consumers instead of competing with each other.

The response of the Roosevelt Administration and the Congress to the intense competition was the Agricultural Adjustment Act of 1933. It was just what the dairy co-ops had in mind. More important, no one noticed or realized at the time the extent of the power being given to dairy co-ops.

While the proposal for the Agricultural Adjustment Act of 1933 had three parts (taxes, allotment plans, and licenses), it was the third part of the Act (licenses) which was most significant to dairy co-ops. The secretary of agriculture was authorized, at his discretion, to issue licenses permitting food processors to engage in the handling of basic agricultural commodities. These licenses were to be issued subject to certain terms and conditions the secretary deemed necessary to eliminate "unfair practices or charges," which would otherwise prevent a price increase on agricultural products to consumers.

Dairy cooperatives were most enthusiastic about the licensing and marketing-agreement concept. They believed the marketing agreements "[C]ould be useful in remedying the two great weaknesses in prevailing cooperative efforts; namely, the failure of [farmers] to give full support to their co-operative organization, and destructive [retail] price cutting on the part of [milk processors]."

Congress subsequently passed the administration's proposal essentially intact, although the section on marketing agreements and licenses was not clearly described in the legislative history and was dealt with only briefly in the Act itself. Dairy co-ops, nevertheless, saw to it that milk was designated a basic commodity coming within the scope of the marketing agreement and license concept.

Congress clearly intended that the marketing agreements enacted by the secretary would cover prices. Given the complaints of dairy farmers in the early 1930s, it was inevitable that such marketing agreements and licenses would evolve into minimum price-fixing devices.

The experience of dairy co-ops with the licensing and marketing-agreement provisions of the Agricultural Adjustment Act of 1933 was only partially successful. Dairy cooperatives liked the concept, and the licenses initially worked as they were supposed to—the price of milk to consumers was kept artificially high. Nevertheless, the free market was proving difficult to kill—competition kept rearing its ugly head.

The experience of the PMA in Chicago is typical. Immediately prior to the enactment of the 1933 act, the PMA had conspired with the Chicago Milk Dealers Association to maintain the retail price of fluid milk in the Chicago market at an artificially high level. Despite their combined efforts, however, the price of milk continued to decline in the Chicago market as competition increased. The PMA and the Chicago milk dealers then negotiated a new agreement to regulate the price for raw milk as well as the price for retail milk in the Chicago area. On the day the Agricultural Adjustment Act became law, the PMA promptly applied to the secretary of agriculture for a marketing license under the Act for the purpose of reinforcing their price-fixing agreement with the Chicago Milk Dealers. The license was immediately issued and became binding on all participants in the Chicago market. It regulated the price of milk paid to dairy farmers, as well as the retail price. Accordingly, nonmembers of the PMA were required by law for the first time to honor the fixed prices which the PMA had conspired with the milk dealers to establish. Within three months, the PMA managed to have the government raise these prices by more than 20 percent.[31]

The Chicago license was clearly designed to limit competition from outlying farms. It established a price mechanism, as well as a system of monetary checkoffs paid to the PMA on its members' milk and to the secretary of agriculture on nonmembers' milk. This scheme thereby prevented nonmembers of the PMA from receiving a higher net pay than PMA members, who had always paid dues for the operation of the PMA. Additionally, all prices for raw milk were FOB at the farm. Hence, a milk processor buying milk from outlying farms was required to pay not only the fixed minimum price but transportation expenses as well. The 1933 Act therefore removed all incentive for purchasing distant supplies of milk that had previously been priced lower than the PMA's price and provided, with the FOB requirement, a positive incentive not to do so.

Licenses substantially similar to the Chicago license were issued by the secretary of agriculture in fourteen other markets during 1933. Still, things were not always as rosy as the dairy cooperatives wished. In Chicago, the Department of Agriculture received complaints of price-cutting shortly after issuing the license. Despite these violations, the legal division of the Department of Agriculture was reluctant to attempt enforcement of the agreements in court. For good reason—they were afraid of losing. Supreme Court decisions on the early legislation of the Roosevelt Administration had declared unconstitutional portions of the National Recovery Act, which had delegated similarly vague powers to the government in the areas of price-fixing and taxation. In addition, the Capper-Volstead Act of 1922 had exempted the formation of agricultural cooperatives from the antitrust laws. It granted no such exemption, however, when dairy cooperatives *and* milk processors conspired to fix prices, as was being done under the licenses issued pursuant to the 1933 act.

As Arthur Schlesinger, Jr., relates in *The Coming of the New Deal,* milk was a "particular area of tension." Chester Davis, the administrator of the Agricultural Adjustment Administration, told the National Emergency Council in February 1934 that his toughest problem had been milk and dairy products. Rexford Tugwell, an original member of FDR's brain trust, specializing in agricultural matters, wrote in his diary that the milk problem disturbed him the most. Tugwell, however, was doing the dairy co-ops' dirty work for them and attempting to shift the responsibility for falling prices onto the large milk processors—ignoring the price-fixing arrangements between dairy cooperatives and large processors for both raw milk and processed milk. Tugwell was joined in this effort by Secretary of Agriculture Henry A. Wallace, who gave a speech in Madison, Wisconsin, in January 1934 attacking what he claimed to be the excessively high profit margins of milk processors in large urban areas.[32]

Given the many problems of enforcing the original AAA's marketing agreements, Congress moved in 1935, at the behest of the dairy cooperatives, to lend more governmental authority to the cooperatives' price-fixing arrangement. The licenses referred to in the 1933 Act were renamed "orders" because the new phrase would "more accurately . . . describe the nature of the regulatory power conferred."[33] To give the informal price-fixing licenses authorized by the 1933 Act the force of law, the 1935 amendments provided that minimum prices to dairy farmers for various classes of milk would be established by the secretary of agriculture for each geographical area in which an "order" had been established and were to be the same for all milk processors.

More important for the future growth in power of dairy cooperatives, the 1935 amendments granted "block voting" authority to cooperatives on behalf of their members, despite the opposition of elements within the Agricultural Adjustment Administration. These amendments provided that a cooperative, in approving the issuance of a federal milk-marketing order, could cast a vote for all of its members despite any individual farmer's wishes to the contrary. Essentially, the 1935 amendments disenfranchised cooperative members and placed a powerful weapon in the hands of the corporate managers of the cooperatives, a weapon which was to be used with ruthless efficiency in the 1960s and 1970s. Illustrative of the dairy cooperatives' power during the Roosevelt Administration is the fact that the procedure of block voting and its disenfranchisement of thousands of American dairy farmers is not discussed in depth by *any* of the Congressional Reports on the 1935 amendments which merely describe the practice.

At the time of the passage of the 1935 amendments, one respected agricultural expert wrote a book in which he offered the following comments on the price fixing and controlled market structure of the Agricultural Adjustment Act (AAA):

> It has many points of similarity with the cartel movement in industry. Its general price theory is built upon the more aggressive elements of the co-operative marketing movement.

> The practice which the cooperatives sought to perfect was not that of collective bargaining for a price but of collective control of the market movement of the commodity in order that certain price objectives might be reached or at least approached. Though the two procedures differ, the goal is the same.[34]

Minor problems arose for the dairy co-ops in 1936, when the Supreme Court declared the AAA unconstitutional. They moved quickly, however, and in 1937, a compliant Congress passed the Agricultural Marketing Agreements Act, which reenacted the marketing-order provisions of the AAA and its 1935 amendments.

The New Deal years were only a beginning for the dairy co-ops. They had not sought the help of the government to protect marginal dairy farmers, who would otherwise be driven out of business. Marginal dairy farmers did, in fact, continue to leave farming in increasing numbers during the depression. Rather, the dairy co-ops had sought the help of the government to protect themselves from the intense competition of those marginal farmers who were undercutting the co-ops' fixed prices in their desperate attempts to keep their farms.

The help given dairy co-ops by the federal government naturally resulted in higher milk prices for consumers. Nevertheless, the impact was not felt on a nationwide basis because federal regulation was largely confined in the 1930s to large urban areas like New York, Chicago, Boston, and Cleveland.

The task of the dairy co-ops since the New Deal has been simple and direct—extend federal regulation of milk throughout the country.

3

The Rise of the Super Cooperatives

Much had been accomplished by the dairy co-ops and their political allies in Congress during the 1930s. They constructed a legal framework for establishing local and regional milk cartels throughout the country. They also institutionalized the earlier system of classified pricing, thereby solving the enforcement problems which plagued earlier dairy cooperatives.

The tactical problem was to build upon this foundation. Aided by the willing bureaucrats of the Department of Agriculture, the dairy co-ops had succeeded by 1940 in extending the coverage of federal milk-marketing orders to 20 percent of all raw milk produced in the country. This coverage grew to 35 percent in twenty-eight separate markets by the end of World War II. The next ten years saw this figure grow to 50 percent of all fluid-grade raw milk. By 1965, it was 70 percent. Four years later, the growth in federal control peaked at 78 percent of all fluid-grade milk, and this is approximately where things stand today.[1]

The dairy industry has undergone significant changes since the 1930s: the character of the participants in the milk industry has changed; marketing practices have become more integrated; and there have been significant advances in technology. But the federal regulatory scheme for milk has not changed—the Department of Agriculture still regulates milk in the same old depression-era way. The reason for this is that it is not an independent body carefully weighing the interest of all parties. It regulates milk solely to raise dairy farmers' incomes.

Until recently, there were few critical economic studies of the cost to consumers of the federal milk-order system. One of the best pre-1970 critical studies was done in 1967 by Dr. Reuben A. Kessel of the University of Chicago. Kessel discovered that federal milk regulation benefits only those dairy farmers who are regulated by federal orders and that such regulation actively works to the detriment of nonregulated dairy farmers. As for consumers, Kessel found that federal orders artificially raise the price of fluid milk and fluid-milk products and lower prices for manufacturing milk products such as cheese and ice cream.[2]

There has been a trend in the postwar era to larger and larger surpluses of raw milk, caused by the federal order system, a trend commented on by Kessel and documented in a 1973 monograph by Dr. Roland Bartlett of the University of Illinois. According to Bartlett, the unnecessary milk surplus in fifteen large federal order markets increased by 300 percent in the twenty years since 1952.[3] The explanation for these surpluses and the sharp increases is relatively simple. Nobel Prize winning economist Milton Friedman is fond of observing that "economists don't know very much. One thing they do know is how to create surpluses and shortages. If you want to create a shortage of a given product, merely establish an artificially low price. If you want to create a surplus for a given product, establish a price at an artificially high level."[4]

Since federal milk regulation was designed in the 1930s to benefit farmers, not consumers, the Department of Agriculture never bothered to do anything other than mandate artificially high prices for farmers. Bartlett did an exhaustive study in the early 1970s of how the federal order price exceeded the costs of shipping fluid-grade milk from the great dairy areas of Wisconsin and Minnesota and concluded that, in 1972 alone, federal order prices were anywhere from $25 million to $45 million higher than justified by market conditions.

More recent studies in 1974 and 1975 have found that excess federal prices for milk are even higher than Bartlett estimated. These studies have also discovered more things wrong with the federal regulation of milk than excessive prices. They were the work of a group of young antitrust economists, most of whom were working with various government agencies other than the USDA—the Justice Department, the Federal Trade Commission, and the President's Council on Wage and Price Stability.

Because they did not work for the USDA and hence had no vested interest in defending its bureaucratic empire, the findings and conclusions of these young economists were highly critical of the status quo at the Department of Agriculture.

Federal milk regulation artificially increases the price to consumers of drinking milk. The result of this artificially high price has been a continuing decrease in consumption of milk, more than a 20 percent decline in the last twenty years. These higher prices have been estimated to add anywhere from $500 million to $800 million annually to the consumers' milk bill.[5]

Federal milk regulation produces disproportionately large benefits for the wealthiest and largest dairy farmers and little if anything for the small family farmer. Specifically, big dairy farmers average over $2,000 a year in federal benefits while small dairy farmers receive only $135—"a deluge of benefits for large dairy farmers and only a trickle . . . for small dairy farmers" according to the Federal Trade Commission's Bureau of Economics.[6]

Federal milk regulation unnecessarily wastes resources, benefitting neither farmers nor consumers. The waste includes such things as overproduction, decreased consumption of fluid milk because of the high prices and the $22 million a year needed to pay for the bureaucrats who run the federal milk system.[7]

Federal milk regulation impedes the movement of milk from one region of the country to another. Where milk markets in the 1920s and 1930s were inherently local in character because of primitive technology, they are now kept local by the government, despite sophisticated technology. With today's transportation system and cooling equipment, milk could regularly be effectively and economically delivered over half the country, were it not for government regulations.[8]

In the face of all these disadvantages, are there any benefits to federal milk regulation? When pressed, the USDA and their dairy co-op allies usually manage to come up with four such benefits: increases in farm income, stabilization of milk prices, maintenance of an "adequate," or "not excessive," milk supply, and improved dairy farmers' bargaining power.[9]

Yet these "benefits" do not fundamentally contradict the findings of the critical studies. The four major benefits of federal milk regulation cannot accurately be described as anything but a consumer rip-off. The classified pricing system of the federal milk-marketing orders could not survive in a free market. The fact is that each one of these alleged "benefits" forceably takes money from consumers and gives most of it to large dairy farmers in a manner which would not prevail in a free market.

Federal milk market orders do increase dairy farm income, particularly large dairy farmers, but at the direct expense of consumers. The annual excess cost is anywhere from $500 to $800 million.[10]

As for price "stabilization," what makes milk so special? Compared to feed grains and hay, dairy prices have always fluctuated less, even in the years before federal regulation. As for maintaining an adequate supply of milk, this is, at best, another bureaucratic obfuscation. Supply cannot be discussed in a vacuum without considering price. Since studies show that consumers purchase less milk at today's high prices than they otherwise would, there is clearly too much milk being produced. As for improving dairy farmers' bargaining power, federal regulation has done that, with a vengeance.

Dairy cooperatives generally acknowledge that federal milk regulation has more than achieved the goals established in the original legislation of the 1930s. It has eliminated what they called "disruptively low milk prices" and provided "long-run price and income stability for dairy farmers." Translated to language consumers can understand, it has killed competition and substituted monopoly pricing.

It would have been reasonable to expect, therefore, that dairy cooperatives, grown safely to maturity within the protective structure of the federal milk cartels, would be content with their legislative successes. During the 1940s and the 1950s, this was the case. Dairy cooperatives operating in federal milk marketing areas were generally satisfied with their cozy cartels and treated the federal price as *the* price for their raw milk. Since the federal price was, by definition, greater than a free-market price, this satisfaction was not unanticipated.

During the 1960s, however, dairy cooperatives simply got greedy. Ignoring the fact that the USDA was fixing a price for them greater than a free-market price, they decided, because of the generally inelastic demand for fluid milk, to see if even higher prices could be gouged from consumers. As early as 1956, the dominant cooperative in Michigan (Michigan Milk Producers Association) began regularly charging dairies a premium price over the federal price and successfully maintained this "over-order" or monopoly premium through 1960.[11]

For most dairy cooperatives in the first half of the 1960s, however, over-order premiums were at best an occasional occurrence, usually brought on in the short season when the monthly federal prices had not adapted quickly enough to reflect supply and demand accurately. Year-round premiums were an often talked of but seldom achieved goal. As in the 1920s, the big problems for most dairy cooperatives were those independent dairy farmers, who perversely insisted upon the right to sell their milk to whomever they chose for whatever price they believed was fair. It is difficult to start a real monopoly when you have people like that around.

Dairy cooperatives reacted to the problems of competition by going back to Congress. After all, Congress had saved them from competition in the 1930s. Maybe it would do so again. If only those independent dairy farmers could be forced to join cooperatives— or at least be forced to pay dues to cooperatives, so that they would gain no economic advantage from selling their milk cheaply—then things would be fine.

In 1964, a coalition was formed by the National Milk Producers Federation, the National Council of Farmer Cooperatives, and the American Farm Bureau to seek enactment of a farmers' collective bargaining bill, which would prohibit milk and other agricultural processors both from refusing to deal with agricultural cooperatives and from soliciting independent farmers to deal directly with processors rather than join a cooperative.

This was the first major legislative effort for the dairy co-ops since the 1930s. What they wanted was nothing less than a collective bargaining bill protecting and encouraging co-ops the same way the National Labor Relations Act protects and encourages labor unions. But this was the 1960s. What they received was far less.

The first version of such a bill was introduced in 1964. Initial hearings were not held on the bill until June, 1966. The original bill was subjected to many revisions and amendments. One such revision, which was strongly opposed by the dairy co-ops, was the deletion of a provision exempting processing cooperatives from coverage by the bill. This revision was opposed by the National Milk Producers Federation because many of the dairy cooperatives which marketed the raw milk of their members also had processing facilities. While the dairy cooperatives wanted to eliminate opposition from independent farmers and milk processors, they did not want to have to play by the same rules themselves when dealing with their own members, merely because they happened to be engaged in processing milk in competition with "proprietary" milk processors.

The opposition of the dairy co-ops to this revision, and many other amendments, which in their view weakened the bill, was to no avail. The final version of the bill (1) clearly applied its anti-discrimination provisions against bargaining cooperatives as well as corporate processors, (2) reaffirmed the right of an individual farmer not to belong to a cooperative, (3) specifically stated that nothing in the Act was intended to prevent farmers and processors "from dealing with one another individually on a direct basis," and (4) provided that nothing in the Act would require a processor "to deal with an association of producers." By far the most important of these changes were the ones which made cooperatives themselves subject to the prohibitions of the bill. What started out as an effort to use the force of the federal government to coerce recalcitrant independent farmers and dairies into line had now been turned into a weapon which independent farmers could use to protect themselves from the cooperatives.[12]

Thereafter, the dairy co-ops abandoned the bill—the Agricultural Fair Practices Act of 1967—and attempted, to no avail, to secure its veto by President Lyndon Johnson.

The Department of Agriculture made no efforts to enforce the new law for more than three years after its passage. A case was finally brought in 1971 against a large Ohio dairy, which received most of its raw milk from independent farmers. The case went to trial in 1973, and the final decision was a devastating defeat for the Department of Agriculture and its dairy cooperative allies. Although the defendant milk processor was given a minor wrist slap, the court held that a milk processor has rights under the law to refuse to deal with a cooperative and to require independent producers to deal directly with it and not through a co-op. Dr. Robert E. Jacobson, a dairy marketing economist at Ohio State University and a long-time apologist for dairy cooperatives, offered the observation in the *Modern Milk Marketer,* that the government had "won the battle," but the milk processor had "won the war."[13]

The significance of the legislative defeat of the dairy co-ops over the bargaining bill cannot be overestimated. They had failed to achieve their primary purpose in seeking the bill's enactment: to have the government eliminate competition from independent dairy farmers.

Most dairy co-ops knew as early as 1965 that favorable bargaining legislation was not going to be passed. The knowledge of their impending defeat had two important consequences, both of which played important roles in the development of what would come to be called the dairy lobby. The first consequence was the decision to attempt to acquire the market power to charge monopoly prices without the help of the government. The second consequence was a decision to attempt to increase their political power in Washington through the deliberate use of campaign contributions. The second consequence would not have had nearly the impact it did on our government in the late 1960s and early 1970s, however, had not the dairy co-ops succeeded beyond their wildest dreams in acquiring monopoly power over milk prices. It was this success of dairy co-ops in empire building which generated such great amounts of additional revenue that allowed them to easily build up huge political war chests.

How was this monopoly power acquired? Many devices and tactics were utilized. Wide-scale cooperative mergers over several market areas were one of the primary methods. Between 1967 and

1970, over one hundred seventy local cooperatives with seventy thousand members merged into four large multimarket regional "super cooperatives."[14]

The largest of these was Associated Milk Producers, Inc. (AMPI), which was formed in 1969 by a consolidation of fourteen Chicago area cooperatives and Milk Producers, Inc. (MPI). MPI, a large southwestern dairy cooperative, was itself the result of a merger in 1967 between six dairy cooperatives located in Kansas, Oklahoma, Texas, and Arkansas. During its first year of operation, it acquired eight other cooperatives in the southwest area. Prior to its 1969 merger with the Chicago area cooperatives, MPI had acquired dairy cooperatives as far north as South Dakota and Minnesota. After the formation of AMPI in 1969, it continued its pattern of growth by merger. In early 1971, it merged with the large Pure Milk Products Cooperative in Wisconsin, giving it a total of forty-five thousand members. By the middle of 1972, AMPI controlled over 75 percent of the raw milk supplied to Chicago, Madison, Indianapolis, Houston, Dallas, San Antonio, Memphis, and Oklahoma City.[15]

The second largest merger resulted in the formation of Mid-American Dairymen, Inc. (Mid-Am) in 1968 from thirty-one cooperatives in Iowa, Kansas, Missouri, and Illinois. In April 1970, Mid-Am added to its size by merging with both Central States Dairy Cooperative in Omaha, Nebraska, and Twin Cities Milk Producers Association in St. Paul, Minnesota. By mid 1971, it had over twenty-four thousand members.[16]

The third largest merger was Dairymen, Inc. (DI), which was originally formed from eight cooperatives and subsequently acquired sixteen additional cooperatives. The original eight cooperatives with members primarily located in Kentucky, Louisiana, Mississippi, Tennessee, and Virginia, merged on September 1, 1968. Since then, DI has added six other cooperatives by merger, and by 1972 had over ten thousand members located in Georgia, Alabama, North Carolina, and Indiana, as well as in the five original states.[17]

The last regional cooperative to be formed was Milk Marketing, Inc. (MMI), originally a 1970 consolidation of four cooperatives in Cleveland, Akron, Toledo, and Pittsburgh. It grew again by merger in 1978, when it acquired the three largest co-ops remaining in Ohio, as well as a large co-op in Indiana. MMI presently has over ten thousand members in Indiana, Ohio, New York, Pennsylvania, West Virginia, and Maryland.[18] The significance of these mergers was summed up by Glenn Lake, then president of the National Milk Producers Federation:

> As a result of these [mergers], seven cooperative[s] can now speak . . . for one-half of the U.S. milk supply . . . in contrast to more than one hundred co-ops speaking for the same volume of milk in the same number of markets five years ago.[19]

According to dairy economists sympathetic to cooperatives, the economic significance of these mergers is that they stop competition and control the supply of milk. In the early 1970s, when Borden and several other milk processors in the Columbus, Ohio area refused to pay what they considered to be excessive raw-milk prices, the dairy cooperative in Columbus (in coordination with other cooperatives with which it had a price-fixing arrangement) decided to teach Borden a lesson by cutting off its milk supply— although not the milk supply of Borden competitors, who had also refused to pay the increased price. Borden naturally went to federal court and quickly secured a court order against this discrimination. What was impressive to dairy economists, however, was less the illegality than the sheer power of it all:

> I was impressed when I read a copy of the temporary restraining order that Borden's secured against Central Ohio Cooperative Milk Producers this fall. Borden's stated, among other things, that they were *unable to get any supplies of milk* in spite of rather intensive solicitation . . . *this was supply control* . . . [emphasis added].[20]

However, market stability over a wide area must be preceded by individual market control, and there are several other techniques and strategies by which dairy cooperatives achieved and maintained monopoly power. These techniques included such complex arrangements as "price alignment," "full supply" contracts, and a "standby pool." The purpose of all the techniques was the same—to eliminate alternate sources of milk. Dairy cooperative managers readily (albeit anonymously) concede all of this. These managers believe that one of the most important objectives of a dairy cooperative is obtaining a price for raw milk which will give its members the highest net return possible. The most significant internal factor affecting the bargaining power of a cooperative is controlling a high percentage of the raw milk marketed in its procurement area, and the two external factors which most enhance the cooperatives' bargaining power are the existence of federal milk orders, and price-fixing arrangements with other cooperatives.[21]

Price alignment, a euphemism for price fixing, is aimed at eliminating price competition among cooperatives in separate markets. Cooperatives do this because when the price differences between various markets are greater than transportation costs, lower priced milk will move into the higher priced markets. This movement can be stopped if the prices in the various markets are "aligned," so that the price differences are the same as transportation costs. Once cooperatives in various markets (or a single cooperative covering many markets) succeed in aligning or fixing prices in this manner, it is difficult for milk processors to find lower priced milk, and cooperative monopoly power is correspondingly strengthened.[22]

Full-supply contracts require a milk processor to buy its entire supply of milk from a single cooperative. Often, such full-supply contracts are offered to a milk processor on a take-it-or-leave-it basis. From the viewpoint of increasing and maintaining market power, the ultimate effect of such full-supply (or "exclusive dealing") contracts is to dry up the market to which independent dairy farmers can ship. To the extent that a cooperative in a given market has all of the major milk processors tied up in full-supply contracts, independent dairy farmers are left with few, if any, dairy customers for their milk. The economic pressure on these farmers to join the cooperative grows as the number of full-supply contracts in a market increases. As the numbers of independent dairy farmers in a market dwindle, they lose viability as an alternate supply source and milk processors become increasingly dependent on a cooperative for their milk. Milk processors who thereafter refuse to enter into full-supply contracts or pay monopoly prices can find their milk supply from a cooperative literally shut off overnight.[23]

The *standby pool* was a direct result of the competitive problem posed by the large supply of surplus Grade-A raw milk which exists in Minnesota and Wisconsin, was unregulated by the federal government, and had no readily available market of fluid-milk processors. The producers of this raw milk in Minnesota and Wisconsin had substantial incentive to seek out fluid-milk markets elsewhere, whenever they could find a price greater than the manufactured-milk-products price being paid by cheese plants in Minnesota and Wisconsin. The presence of such a large alternate supply of unregulated Grade-A milk effectively placed a ceiling on the monopoly prices which could be charged by the large regional cooperatives.

The standby pool was intended to meet this problem. It was a "federated cooperative" whose members included the four super cooperatives, as well as fifteen other cooperatives. Each member cooperative in the standby pool paid monthly assessments (at one time, two cents for every hundred pounds of milk) on all fluid milk sold by each member in their respective regional markets. The money generated was used by the standby pool to pay Grade-A milk producers in Minnesota and Wisconsin *not* to ship their milk into other regions without approval. In plain words, these Minnesota and Wisconsin producers were:

> bribed to refrain from spoiling the markets of [standby pool member] cooperatives which have successfully demanded [monopoly] prices.[24]

While tactics such as these enabled the super cooperatives to achieve monopoly or near-monopoly positions, even government economists agree this could not have been accomplished without the underlying federal regulatory system itself. In a study for the USDA prepared by Alden Manchester of the USDA's Marketing Economics Division, he concludes that "Federal and State orders provide the institutional framework within which a cooperative can bargain successfully for prices higher than those in the orders."[25] Other economists made derisive references to "the passive willingness of the USDA to allow cooperatives to dictate Federal Order provisions."[26] This passivity becomes all the more important in terms of the USDA's complicity in the growth of the power of super cooperatives. The secretary of agriculture has the authority under the Capper-Volstead Act to take action against agricultural cooperatives whose monopoly power enables them to "unduly enhance" prices. But that authority, on the books for over fifty years, has never been exercised.

What were these super cooperatives really after in the late 1960s and early 1970s? Pretrial testimony in the government's antitrust suit against AMPI by David Parr, a high level AMPI executive, offers a clue. AMPI's eventual goal, said Parr, was a single nation-wide dairy cooperative with a "total marketing concept"—production, promotion, processing, and political lobbying:

> *Question.* Is that all dairy farmers?
> *Answer.* Yes.
> *Question.* Including nonmembers?
> *Answer.* If it was up to me, there wouldn't be any.[27]

What about consumers in this grand scheme? "You never know what consumers are willing to pay," said Parr, "until you ask for it."[28]

None of the super cooperatives had any inhibitions about asking. They literally charged as much as, and frequently more than, the market would bear. They did this in defiance of the USDA, which is required by law to police "undue price enhancement" by cooperatives.

In February 1976, the National Consumers Congress filed a petition with Agriculture Secretary Earl Butz requesting that he institute a proceeding under the Capper-Volstead Act against all dairy cooperatives in every Federal Order Area which had established monopoly premiums for their milk. In December of 1976, after a brief five month "investigation," the Agriculture Department gravely announced that it had looked high and low and could not find any cooperative charging monopoly prices anywhere. But then again, maybe they were not looking in the right places. Indeed, they probably were not even looking at all. For starters, they might have talked to the Justice Department, which had issued its own massive six hundred page *Study on Milk Marketing* which found that monopoly premiums across the country cost consumers anywhere from $210 to $260 million a year more than was justified by market conditions.[29]

Better still, they might have consulted American consumers generally who, unorganized and unaided by consumer groups, have been quietly engaged in a massive fifteen-year-long boycott of whole milk. In 1964, the average consumer drank thirty-eight gallons of milk a year. In 1978, the figure had dropped by more than a third—to twenty-four gallons a year.

A truly free market would react to such a massive boycott by lowering prices. Look what happened a few years back when bad crop conditions caused the price of sugar to skyrocket. A semiorganized consumer boycott ensued, crops improved, and prices dropped. Not so with milk. There is no free market. The spontaneous consumer boycott has been met with massive indifference by both the bureaucracy at the USDA and the super cooperatives. Why? Because the super cooperatives—the dairy lobby—believe that politicians are more important to their economic survival than consumers.

All of which brings us back to the second consequence of the defeat suffered by the dairy cooperatives on their bargaining bill back in 1967: a deliberately planned increase in their political influence in Washington. The dairy co-ops' efforts at acquiring market power had been successful in terms of generating additional money that could be used to reward their friends in Washington and at least purchase "access" to argue their case before those who were neutral or indifferent.

Pursuant to this policy, the dairy co-ops became heavily involved in the 1968 presidential campaign. Most of their involvement was in the form of illegal campaign contributions to President Lyndon Johnson and then Vice-President Hubert Humphrey. After that campaign, the dairy co-ops started to approach the question of campaign contributions in a systematic way. The three largest dairy co-ops—AMPI, DI and Mid-Am—all created political "trusts" in 1969 and 1970.

Associated Milk Producers, Inc. (AMPI), the largest and initially the most politically active dairy co-op, created a political trust called TAPE. TAPE, which was headquartered along with AMPI in San Antonio, Texas, was the acronym for Trust for Agricultural Political Education. The basic premise of TAPE was that its donors—dairy farmers who were members of the cooperative and individual AMPI employees—would make contributions by a checkoff formula of just under $100 per year, thus avoiding public reporting of the identity of contributors under the then applicable Corrupt Practices Act of 1925. TAPE was succeeded in 1972 by the Committee for Thorough Agricultural Political Education ("CTAPE"), which was formed in order to "involve more persons in the decision-making process" after passage of the Federal Elections Campaign Act of 1971.[30]

Dairymen, Inc. (DI), not to be outdone, formed the Trust for Special Political Agricultural Community Education (SPACE). AMPI consulted with DI in the formation of SPACE, and it was created in the same fashion as TAPE, with checkoffs from donors just under $100. DI had only ten thousand members (compared to the forty-thousand members of AMPI), and SPACE was therefore a smaller political trust.[31]

The third major co-op, Mid-America Dairymen, Inc. (Mid-Am), covers portions of the midwest and is headquartered in Springfield, Missouri. With twenty thousand members, it is the second largest dairy cooperative. In 1970, after consultation with AMPI, Mid-Am formed ADEPT, an acronym for the Agricultural and Dairy Educational Political Trust.[32]

The persons primarily responsible for the formation of TAPE and subsequently instrumental in the formations of SPACE and ADEPT were Harold Nelson, David Parr, and Jake Jacobson. More than anyone else, it was their perception of the appeal to politicians of mass contributions from special interest groups which made possible the milk-fund scandals of the 1970s.

Harold Nelson was manager of the Texas Milk Producers Federation in 1959. The Federation consisted of the South Texas Producers Association from Houston, several producers associations in Corpus Christi, the North Texas Producers Association of Arlington, Texas, the Mid-Texas Producers Association of Austin, and the Producers Association of San Antonio. In 1967, the Federation ceased operations when Milk Producers, Inc. (MPI) was formed. Nelson was made General Manager of MPI and later became General Manager of AMPI upon its creation in 1969. He was, in effect, its chief executive officer.

From August, 1953 until September, 1967, David Parr was manager of the Central Arkansas Milk Producers Association. He was a division manager of MPI until 1969, when he joined the staff of AMPI, where he served until 1972. Parr was actively involved in the politics, promotions, and consolidations which resulted in AMPI's formation.[33]

While Nelson and Parr were primarily responsible for the formation of TAPE, Jake Jacobsen was the chief lawyer advising them. He was the senior partner of Jacobsen and Long, a law firm in Austin, Texas, and also a partner in Semer and Jacobsen, a small Washington, D.C., firm formed in 1968. Jacobsen first met Nelson and Parr at the White House, where he worked as a close aide to Lyndon Johnson. The firm of Semer and Jacobsen was retained in 1968 by AMPI and, according to Jacobsen, he "primarily gave advice in the political area" and discussed in detail with Nelson the trust fund setup which led to the formation of TAPE. His advice involved both the mechanics of setting up such a trust fund and its feasibility.[34]

The mechanics and feasibility were remarkably simple. The three trusts—TAPE, SPACE and ADEPT—controlled between them almost eighty thousand dairy farmers, who received checks for their milk every month from the parent co-op. A minor check-off of $8.25 a month would account for an annual political contribution from each farmer to the trust of $99. Multiply this by seventy thousand farmers, and one can see a staggering potential of $6.9 million a year which could be earmarked for political contributions.

What were the purposes for amassing such huge sums, and how were they to be used?

The purposes, at least to begin with, were superficially modest. The executive branch of the government regularly establishes price-support levels at which it will purchase all surplus milk for which there are no other buyers—it functions as a buyer of last resort. It was important to the dairy co-ops to have this price-support level as high as possible because it plays an important role in establishing the overall price for raw milk throughout the federal milk-marketing system.

Similarly, the president regularly establishes import quotas for foreign dairy products. It was important to the dairy co-ops to keep foreign dairy imports low because they too played an important role in establishing the overall price for milk in the federal milk marketing system.

Later, as the notoriety of the dairy co-ops grew, and the questionable nature and illegal tactics involved in their rapid rise to power came to light, their political activity would expand to include influencing government antitrust policy—both prosecutorial and legislative—as well as defending the very existence of the federal milk-marketing program established in the depression.

Consistent with their early goals, the initial contributions of the dairy co-ops were focused on the presidential level—presidents and candidates for president. Back in the pre-Watergate reform era, a presidential race was where the action was and where it was possible to have the most impact on policy. Not that the dairy co-ops focused on presidential campaigns to the exclusion of Congress. Far from it. They always kept their Congressional fences mended. But Congress in those pre-Watergate days was mainly a back-up to be used to prod and pressure the executive branch when it failed to move as quickly or as far as the dairy co-ops wished.

The administration of Lyndon Johnson and the campaigns of his vice-president, Hubert Humphrey, are a good place to begin an analysis of the present day dairy lobby. For one thing, 1968 was the dairy lobby's first presidential campaign. Dairy co-ops had not played a significant role in 1964 because their mergers and consolidations of economic power were yet to occur. The lessons the dairy lobby learned in 1968—not the least of which was never put all your eggs in one basket—were to play a prominent part in the Nixon years. Another reason is that in 1968 the dairy co-ops had no vehicle or formal structure for making large campaign contributions legally—their political trust funds were yet to be formed. Their political efforts in 1968 were, therefore, largely illegal and present something of a contrast to the mixture of legal/illegal contributions in succeeding campaigns.

A final reason for beginning our analysis with LBJ and Hubert Humphrey is that they lend a bipartisan cast to a study which necessarily forcuses to a large extent upon Richard Nixon and Watergate. Bipartisanship is important to the analysis, however, because it helps to focus on the dairy lobby's approach to politics. The dairy lobby does not favor one party over the other. It goes where the power is. If the dairy lobby has given more money to Democratic senators and congressmen than to Republicans—and it has—that is only because the Democrats are the majority party in Congress. If the dairy lobby gave more, legally and illegally, to Richard Nixon than it did to various Democratic presidential candidates—and it did—that was because Nixon was the president and in the 1972 election seemed likely to stay that way. When the Watergate scandals and the 1974 campaign reform measures ended for the foreseeable future the era of huge presidential campaign contributions, the action shifted to Congress. Studying those results clearly reveals the dairy lobby's approach to politics. The money of the dairy lobby flowed to incumbents by an eight to one ratio, regardless of party or political philosophy. In other words the dairy lobby supported the persons most likely to be elected, i.e. incumbents. And if a race had no incumbent, the dairy lobby often gave to both sides.

In 1967, all this was in the future. Harold Nelson had just presided over the formation of MPI out of six dairy co-ops located in Kansas, Oklahoma, Texas, and Arkansas; MPI was headquartered in San Antonio; Nelson was its general manager; and his old friend, Lyndon Johnson, was the president.

Part 2

The Abuse of Power I: Milk and Presidential Politics

These dairymen are organized; they're adamant, they're militant.
... And they, they're massing an enormous amount of money that
they're going to put into political activities, very frankly.

Secretary of the Treasury John Connally
March 23, 1971

4

The Democrats: Lyndon Johnson, Hubert Humphrey, and the Ways and Means of Wilbur Mills

Back in 1967, Lyndon Johnson was still running for re-election. The 1968 Tet offensive, his near-upset by Eugene McCarthy in the New Hampshire primary and—most dreaded of all—Robert Kennedy's campaign for the presidency, were all still in LBJ's future. Someone in the White House, during the late summer of that year, had the bright idea of publishing a book showing the President in a noble light—a collection of his messages to Congress titled, *No Retreat from Tomorrow*.[1]

The White House thereafter formed a "Salute to the President" committee and had it contract with Harlowe Typography, Inc. of Washington, D.C., for the typesetting ($7,910.51), with Artwork Unlimited, Inc. of Washington, D.C., for graphics ($5,900), and with McGregor and Werner, Inc. of Washington, D.C. for printing, binding, and mailing to friends and potential contributors 26,184 copies ($90,711.07).

The books were ready by January of 1968, and invoices from the three companies were submitted by March of 1968. Unfortunately, by March 21, 1968, Lyndon Johnson was in full retreat from tomorrow—he had dropped out as a candidate for re-election. Still, the work was done; the bills were there; and they had to be paid. The Democratic National Committee was not particularly enthusiastic about paying the bills for a lame-duck president. They had their own presidential campaign—Hubert Humphrey's they hoped—to finance. Besides, the "Salute to the President Committee," which ordered the books, had not been sponsored by the Democratic National Committee. It had worked out of the White House. Let the White House pay for its own bills.

LBJ knew just where to go to get the bills paid. Johnson believed that a large group of Texas dairy cooperatives had not fulfilled a $250,000 "commitment" to his 1964 campaign for president. Since LBJ as a senator and president had long been a friend of the dairy lobby and had consistently supported dairy interests in such areas as price supports and import quotas, he arranged to have all of the invoices brought to the attention of Harold Nelson, then general manager of the recently formed Milk Producers, Inc. (MPI), the predecessor of AMPI. Fortunately for the dairy lobby, March of each year is when the secretary of agriculture adjusts the government price-support level for surplus milk, and on March 29, the Johnson Administration increased milk price-supports by 7 percent, even though the cost-of-living increase during the past year had only been 2.8 percent.

The dairy lobby thereafter picked up the entire cost of the publication. MPI paid for the typesetting and graphics directly. Nelson then had McGregor and Werner, Inc. break down the $90,711.07 invoice for printing into three invoices, one for $30,250, another for $28,500 and a third for $31,961.07. These invoices were distributed respectively to the Central Arkansas Milk Producers Association, The North Texas Producers Association, and the MPI office in Dallas. Since the dairy lobby had not yet formed its political trust funds, all three invoices were promptly paid with corporate funds.

These campaign contributions from corporate funds were almost certainly illegal. Even though Johnson was no longer a candidate when the invoices were paid, he clearly had been a candidate when the books were ordered, printed, and distributed. All of the invoices were rendered *before* March 21, 1968, when Johnson withdrew from the presidential race.

The timing of the payments, illegal or not, is even more questionable. Coming as they did after a presidential decision on price supports, the payments were to have an ominous parallel in the Nixon Administration, with far-reaching consequences for the careers of Richard Nixon and his treasury secretary, John Connally.

The dairy lobby's gratitude for LBJ's 1968 price-support decision, however, did not stop at repaying $100,000 of LBJ's campaign debts. Harold Nelson and MPI's successor, AMPI, continued to be kind to LBJ after he left office. To supplement his retirement, AMPI agreed to lease a Beechcraft King Air from one of LBJ's companies.

The original lease ran through June, 1972 and provided for a base rent of $73,500 a year plus $160 an hour after the first thirty-five hours of use each month. AMPI entered into a new lease with LBJ in January, 1972, this time for four years, rental for which ran to $94,000 a year.[2] There was a catch with both leases. On the first lease, AMPI had to leave the plane at the Johnson City, Texas airport and return it there after each flight. The second lease required the plane to be kept at the LBJ ranch itself.

When all this came to light in March of 1974, Donald Thomas, a spokesman for the LBJ Corporation, described the lease deal as a "bargain" for AMPI.[3] AMPI, however, did not quite see it that way and tried to rid itself of this "bargain" in 1972, when the lease was up for renewal, but did not act quickly enough. Harold Nelson extended it in January, 1972, as one of his last acts in office. When AMPI then tried to peddle its lease rights to other firms, there were no takers.

The ties between Johnson and the AMPI management, after Harold Nelson resigned in early 1972, remained close. In the fall of 1972, new AMPI General Manager George Mehren, who had been an assistant secretary of agriculture in the Johnson Administration, flew to the LBJ ranch to seek advice from his old boss. Mehren told Johnson of the continuing pressure AMPI was receiving from the Nixon Administration for additional campaign contributions. Mehren had met that morning with Lee Nunn, vice chairman of the Finance Committee to Re-Elect the President. Nunn had just asked Mehren for $650,000. Nunn told him that such a contribution could never be a *quid pro quo* but that "it is correct that the president does remember his friends who helped him." Nunn had also claimed that the Nixon campaign was $10 million in debt. Upon hearing this retold by Mehren, Johnson was incredulous. "Do you really believe that?" he asked. "No," Mehren responded, you "didn't ask me what I believed. [You] asked me what Mr. Nunn had said." During the rest of their meeting, Mehren and Johnson discussed how AMPI should respond to the Nunn solicitation. Johnson advised and Mehren eventually agreed that AMPI should ignore Nunn and make identical contributions to the congressional campaign committees of both parties.[4]

Mehren hardly needed to make a special flight to the LBJ ranch to receive such routine advice. It is quite likely, therefore, that Mehren also raised with LBJ the possibility of letting AMPI out of the new four-year airplane lease. If Mehren did raise this question, it helps to explain what else LBJ told Mehren at that meeting. According to Mehren, Johnson said that milk producers had once promised to deliver $250,000 to him, a promise, complained Johnson, which was never kept. It is possible this was LBJ's way of gently suggesting that, given the failure of predecessors to come through with the promised $250,000, it was slightly ungracious for AMPI to ask to be released from its $94,000 a year "bargain" rental for LBJ's airplane.[5]

Because of LBJ's well-deserved reputation as a wheeler and dealer, his unsavory relations with the dairy lobby, both during and after his years in the White House, may no longer seem shocking or surprising.

Hubert Humphrey, however, is another story.

Hubert Humphrey was considered by many as a political folk-hero, particularly in the years after his failure to gain the 1972 Democratic presidential nomination. To those who believed in him, he conveyed to the end the image of the ebullient, stout-hearted fighter for the rights of the down-trodden, a public man who wanted to use government for what he thought was the common good. Humphrey's own autobiography illustrates his self-perception:

> Democrats seem to love government, while, I suspect, high-level Republicans too often really do not. For many of them service in Washington is often nothing more than a break between two jobs in private industry or banking or law, and the art of

government is itself less appealing, less exciting than it is to Democrats.[6]

Nevertheless, in order to consummate his love for governing others, Humphrey always needed to be elected, and it was here where he had problems. Most observers readily acknowledge Humphrey's genuine love for oratory and meeting people. But successful political campaigns involve much more than shaking hands and giving speeches. He needed organization for one thing, and money for another—lots of it. *This* was the side of politics which gave Humphrey the most personal pain. He simply did not like raising money for his many campaigns; he did not like the small, intimate meetings with wealthy contributors where he had to showcase himself and, in his view, literally beg for contributions. He found it demeaning and thought the entire process distasteful and far removed from the compassionate ends of governing other people which he envisioned and enjoyed.

It is not known whether Humphrey's fund raising relationships with the dairy lobby were as distasteful to him as all the rest. Humphrey in his later years alternately stonewalled or denied knowledge of the details of his dairy lobby contributions. Humphrey's ties to the dairy lobby extended over a period of many years, including his 1968 campaign for the presidency against Richard Nixon, his 1970 bid to return to the Senate from Minnesota, and his abortive 1972 presidential campaign. All of these campaigns involved the receipt of illegal corporate campaign contributions. While Humphrey was not unique in this respect, there still lingers today a nagging suspicion that his position in the Democratic party and his almost revered stature helped him to escape substantially all public taint surrounding the Senate investigation of his ties to the dairy lobby.

The first Humphrey/dairy lobby connection occurred in 1968, when he and Edmund Muskie were attempting to retain the presidency for the Democratic party. In 1968, the dairy co-ops, whose management was traditionally allied with the Democratic party, went all out for Hubert Humphrey, since he had been a supporter of dairy interests in the U.S. Senate since 1949 and had never voted against an increase in government milk-price supports, unless he believed the increase too small.

Dairy lobby contributions in the 1968 campaign were virtually all illegal, since the various political action committees were yet to be formed. Thus, most 1968 contributions were made by the cooperatives with corporate funds. To obscure and cover up the illegal corporate contributions, the dairy lobby used a number of laundering schemes. For example, trusted milk cooperative attorneys would make personal political contributions and then be reimbursed by billing the cooperatives for nonexistent legal services. Similarly, employees of the co-ops would make personal contributions and, through the art of creative expense accounting, would be reimbursed with corporate funds.

This policy, endorsed and supported at MPI and AMPI by Harold Nelson, was very costly, since the attorneys would have to bill for fees almost twice the amount of the contribution to recover the cost of the contribution *plus* the money needed to pay their income taxes on the extra billing. One such attorney, Stuart Russell, who maintained that AMPI directors were "generally aware" of these practices, sent *excess* billings of $200,000 to AMPI over a four-year period.

The initial illegal contribution to Hubert Humphrey was a nine-month loan of the full-time services of AMPI's resident political expert, Bob Lilly. Lilly had come from a political background, having represented the State Farm Bureau of Texas before the Texas State Legislature during the 60s. He had been with the dairy industry since 1965, when he worked for the North Texas Producers Association in Arlington, Texas. After AMPI was formed, he became assistant to General Manager Harold Nelson and was primarily responsible for legislative activity. Lilly's first activities consisted of efforts "to sew up delegates for the nomination" of Humphrey. Later in 1968, Lilly worked intensively on the states of Alaska, Hawaii, Indiana, Iowa, Washington, and West Virginia for Humphrey and was state coordinator for Alaska and West Virginia. These activities were all arranged and approved by Nelson, who saw to it that Lilly's salary and all of his expenses were paid by the co-op while he worked for Humphrey.[7]

By Lilly's own estimate, he also spent more than $40,000 in 1968 for political contributions. To repay him for these expenditures, AMPI created a system whereby he received a series of bonuses and salary advances to reimburse him for the contributions. This money was placed in separate checking accounts, from which Lilly wrote checks to disburse money.

AMPI also helped Humphrey in 1968 through the illegal device of $1,000 advances to various employees. AMPI bookkeepers charged these $1,000 advances, to such items as "professional services, director's expenses and travel." In each case, the employee subsequently made a contribution of $1,000 to the Humphrey campaign.[8]

Other illegal conduits to the 1968 Humphrey campaign from AMPI included Holiday Camps of Arkansas, Inc. and the Arkansas Action Committee for Rural Electrification. Holiday Camps was incorporated in 1968 by Oscar Alagood of Little Rock, a former Arkansas state senator, who had been for several years a friend of AMPI executives David Parr and Keiffer Howard (of AMPI's Arkansas Division). At the request of Parr, Alagood made a personal $6,000 contribution to the Humphrey campaign. On the same day, Parr issued a $7,000 corporate check to Holiday Camps. Inc., to reimburse Alagood for his contribution to Humphrey. The extra $1,000 was a thank-you to Alagood for his troubles serving as a conduit, a "loan" which six years later had not been repaid. Parr and Howard similarly had the Arkansas Action Committee for Rural Electrification contribute $20,000 to Humphrey and then reimbursed the Committee with AMPI's corporate funds. For their troubles on behalf of Humphrey, Dave Parr and Keiffer Howard were convicted, fined, and placed on probation.

By comparison, the dairy lobby was positively ungenerous with the Republicans in general and Richard Nixon in particular in the 1968 presidential campaign. The most that AMPI did was have Harold Nelson personally purchase $5,000 of tickets to a GOP victory dinner in May—money which was illegally reimbursed to him from corporate funds and charged to "advertising and public relations."[9] It was the one-sided and illegal support for Humphrey, however, which laid the groundwork for the dairy lobby's subsequent corrupting contributions to the newly-elected Nixon. It had a lot of catching up to do.

1970 saw the return of Hubert Humphrey to public life (after a brief respite as a professor at the University of Minnesota) when he decided to run again for the U.S. Senate. Humphrey chose as his 1970 campaign manager Jack Chestnut, a bright (Harvard Law School) attorney practicing in Minneapolis, who was to become one of the more tragic figures in the political scandals of the dairy lobby: his previously promising legal career was to be tarnished by criminal indictment, prosecution, and conviction in connection with his Humphrey campaign role.

As Humphrey's campaign manager, Chestnut made an agreement with the New York based advertising firm of Lennen and Newell, Inc. for advertising services to be delivered to the Humphrey campaign at a total cost of $72,000, payable to Lennen and Newell in monthly installments over the six-month course of the campaign. Some time in March or April, 1970, Chestnut telephoned Lennen and Newell account executive Barry Nova to tell him that the financial backers of the Humphrey campaign were not performing as hoped and that funds were not available to pay for the required advertising services. According to Nova, Chestnut asked that the next monthly bill be directed to "American Milk Producers" (Nova's unprompted recollection of the term used by Chestnut).[10] Lennen and Newell in fact billed Chestnut for its work on the Humphrey campaign to "Associated Milk Producers, Inc., c/o Bob Lilly, New Ulm, Minnesota."

When the bill was received, one of AMPI's New Ulm employees phoned Lilly in complete puzzlement about the $12,000 bill from such a big advertising firm. Lilly, who claims he was also in the dark at that point, merely asked that the bill be forwarded to him in San Antonio. Lilly then asked Harold Nelson for an explanation of the bill, and Nelson told him that the bill was from a New York advertising firm and, that while it appeared to be a bill for services rendered to AMPI, it was actually for services performed by Lennen and Newell for the Humphrey Senatorial Campaign. Nelson approved the payment and told Lilly to talk to Chestnut to arrange payment.[11]

Lilly talked with Chestnut several times in the days that followed and expressed his displeasure at the bill being sent to the New Ulm office, where no one was aware of AMPI's political activity. They proceeded to work out a different arrangement whereby

Chestnut would send the bills directly to Lilly, who would return the checks to Chestnut for transmittal to Lennen and Newell.

In May, 1970, Chestnut sent four invoices from Lennen and Newell to Lilly, each in the amount of $3,000. On the first of June, Lilly sent to Chestnut two $6,000 corporate checks from AMPI payable to Lennen and Newell which Chestnut forwarded to New York. For his role in this scheme, Chestnut was indicted and convicted in 1974 for soliciting illegal corporate political contributions.

In commenting on the illegal contributions when they first came to light, Humphrey refused to speak directly to reporters. Instead, he plead ignorance through a "spokesman":

> I have no knowledge of these transactions . . . an organization as large as AMPI should have had the kind of legal counsel that would have prevented these types of transactions.[12]

Humphrey apparently did not believe it was his campaign organization's responsibility to prevent "these types of transactions" because they happened again in 1972 in his campaign for the presidential nomination. During his 1970 Senatorial campaign, Humphrey had made use of a new firm, Valentine, Sherman and Associates ("VSA"), which compiled computerized voting lists for use in mass mailings. One of VSA's principals, Norman Sherman, had been Humphrey's press secretary when he was vice-president. VSA's lawyer was Jack Chestnut, again serving as Humphrey's campaign manager in 1972. VSA was active in Humphrey's 1972 presidential campaign as well and billed the Humphrey organization $270,000 including $200,000 for computer-mailing services in Nebraska, Oregon, Florida, and Maryland. Twenty-five thousand dollars of this was illegally paid directly from AMPI corporate funds with Chestnut's knowledge.[13] Other than the illegal $25,000 from AMPI, the 1972 Humphrey presidential effort received only $17,225 directly from dairy committees.

As will be examined in a later chapter, Hubert Humphrey willingly went along with the dairy lobby's efforts in early 1971 to pressure the Nixon Administration publicly and privately into reversing an unfavorable price-support decision. Yet Hubert Humphrey, who was the featured speaker at the annual AMPI convention in June of 1973, was pronounced by the Senate Watergate Committee to be totally uninvolved with any illegal activity vis-à-vis AMPI and clear of suspicion that his vote had ever been bought. The clean bill of goods, however, came about without Humphrey's cooperation.

Humphrey's openness with the committee can be charitably described as less than perfect. Humphrey's office provided copies of his schedules on a daily basis to the Senate Watergate Committee upon request, but failed to honor any other requests.[14] On January 24, 1974, Senator Sam Ervin wrote asking Humphrey to meet with members of the Committee staff "concerning various allegations concerning the employment of corporate funds by Associated Milk Producers, Inc. and others."[15] Humphrey ignored the letter so Ervin sent another on February 7 as a "follow-up" in which he specified that he believed it "necessary for a committee member to speak with you." Humphrey responded to both letters on February 20 and "referencing to corporate funds paid by VSA to AMPI," stated:

> Let me say that neither at the time of the alleged transaction nor now do I have any knowledge concerning this particular matter . . . Because I know nothing about the transaction and *have no records* in my files relating to it, I see no point in *inconveniencing* any member of your Committee to meet with me [emphasis added].[16]

What Humphrey really meant was that virtually all of his campaign financial records for the period prior to April 7, 1972 had been destroyed. While Humphrey had no legal obligation to maintain the records, the Senate Watergate Committee dryly noted that "such wholesale destruction necessarily raises a question of motive and propriety" and quoted with approval Senator Sam Ervin's comments to Nixon fund-raiser Maurice Stans, who had similarly destroyed all of Nixon's pre-April 7, 1972, campaign financial records:

> . . . you decided that the right of the contributors to have their contributions concealed was superior to the right of the American citizens to know who was making contributions to influence the election of the President of the United States . . . do you not think that men who have been honored by the American people

. . . ought to have their course of action guided by ethical principles which are superior to the minimum requirements of the
criminal laws?
. . . do you think that men who exercise great political power
. . . ought to disregard ethical principles and say they have fulfilled
their full duty to the American people as long as they keep on
the windy side of the law?[17]

Humphrey's refusal to meet with the committee spared him
the embarrassment of responding to such questions. As a result,
Humphrey's stonewalling tactics were successful. He never had to
account for any role he may have played in his two successive
campaigns which were financed by illegal dairy funds. Jack Chestnut, his closest adviser in those campaigns, was less fortunate.

Humphrey's public silence on his brush with Watergate continued in his 1976 autobiography, *The Education of a Public Man,*
in which he took great pains to establish a considerable distance
between himself and his campaign manager, Jack Chestnut. In a
manner reminiscent of Richard Nixon's disclaiming responsibility
for what *his* campaign manager, John Mitchell, had done during
Watergate, Humphrey professed hardly to have known the single
most important man in his last two campaigns. The only reference
Humphrey made to Chestnut was a condescending little anecdote
about Chestnut's role as an advance man back in 1968, when his
father had died on election day but Chestnut loyally refused to
leave Humphrey's side until arranging plans for the relatively
minor task of entertaining the press.

Whether Chestnut deserved something better from his former
boss, clearly the illegal contributions probably would not have been
made at all, were it not for Richard Nixon and his top aides' version
of political hardball.

Unlike those of 1968, the dairy lobby contributions to Humphrey
in 1970 and 1972 did not have to be illegal. All three super co-ops
had by that time formed the political action committees to which
their dairy farmer members voluntarily contributed. Campaign
funds from these sources to Humphrey in 1970 and 1972 would
have been legal.

The fact that no legal contributions were forthcoming to Humphrey, however, says a lot about the dairy lobby and its approach

to politics. In 1970, Humphrey was not in office and hence was in no position to help the dairy lobby. Even by 1972, Humphrey had not regained sufficient stature to be considered a leading candidate for the Democratic presidential nomination. For, at the same time, illegal—and hence unreported—contributions were flowing to Humphrey, far greater contributions, legal and illegal, were being delivered by the dairy lobby to Richard Nixon. Nixon now possessed the power over such important dairy matters as import quotas and price supports, and the dairy lobby was eager, as will be examined in detail in subsequent chapters, to mend its fences with the Republicans to atone for its near-exclusive support of Humphrey in 1968.

The problem for the dairy lobby was that Nixon and his top political aides—most notably Charles Colson—disliked special interests playing both sides of the street and would certainly have frowned upon any dairy lobby contributions to the campaign of Nixon's 1968 rival.

The solution of the dairy lobby to this dilemma was characteristic of their approach to politics. If it cannot be done legally, then do it the other way. Humphrey had not objected to the illegal corporate contributions in 1968, and the dairy lobby correctly reasoned there would be no problem with similarly illegal contributions in 1970 and 1972. Indeed, but for Richard Nixon and Watergate, no one would ever have found out about Humphrey's illegal financing, and even this relatively small stain would not have marred Hubert Humphrey's memory.

There was someone else far more important to the dairy lobby in the campaign for the 1972 Democratic presidential nomination than Hubert Humphrey. The dairy lobby had shrewdly calculated early on that the man who lost to Richard Nixon in 1968 would not be given a second chance in 1972. And the dairy lobby was not about to put all its eggs in Nixon's basket.

The Democratic basket chosen by the dairy lobby for the 1972 presidential nomination belonged to Wilbur Mills, the chairman of the prestigious House Ways and Means Committee. Considered by many knowledgeable observers to be the most powerful person on Capitol Hill in 1971, Mills was literally the dairy lobby's own candidate for president. It bankrolled almost 40 percent of the $500,000 Mills spent on his unsuccessful campaign.

As with Humphrey, and for the same reasons, a large part of the dairy lobby's contributions to Mills was illegal. Mills did not object at the time, but the tragic consequences for him, in the aftermath of Watergate, were to end his remarkable thirty-eight year career in Washington.

In 1971, less than four years before his fall from power, Wilbur Mills was a powerful man because of the congressional committee he chaired and the type of chairman he was. Ways and Means is the oldest and most eminent committee in Congress, originally established in 1789. Ways and Means has power over all taxes, including not only income taxes but Social Security, Medicare, and tariffs. What makes it even more powerful is the constitutional requirement that all tax measures originate in the House of Representatives. This, in the congressional scheme of things, gives it more power than its Senate counterpart, the Finance Committee, which may pass on revenue measures only after the Ways and Means Committee has devised them. And if that were not enough power, the Democratic members of the Ways and Means Committee for years constituted the "committee on committees," which handed out committee assignments to all the Democratic representatives in the House.

Wilbur Mills was born in Kensett, Arkansas, a small farm community fifty miles from Little Rock, on May 24, 1909. After graduating from a small Methodist college in Arkansas, Mills went to Harvard Law School and graduated in 1933. He returned to Arkansas and, barely a year out of law school, successfully ran for County Judge of White County, ousting a long-time incumbent. Four years later, he was elected to Congress. He was only twenty-nine.[18] After 1944, Mills never had an opponent for re-election until his last campaign in 1974.[19]

In many ways, Mills was a protege of that legendary speaker of the House, Sam Rayburn, who, early on in Mills' career, recognized his superior abilities, his safe district, his moderate-to-conservative political leanings, and decided to place him on the Ways and Means Committee. Through the inexorable workings of the seniority system which then prevailed in Congress, Mills in 1958, at the relatively young age (for the House) of forty-eight, became Chairman of the Ways and Means Committee. From then until his fall in 1974, Mills' power in Congress over matters under his committee's jurisdiction was almost absolute, a fact resulting from Mills' sheer expertise on the complicated matters coming before his committee. One Republican committee member commented in the early 1960s on Mills' expertise and hard work:

> Mills is pre-eminent. He's a real student. I don't know what the Democrats will do when he goes. He was the real leader before he became chairman because he studied so much. [His] leadership depends on hard work and knowledge.
>
> He knows the tax code inside and out and he knows what Ways and Means has done for the last twenty years. He can cite and does cite section after section of the code. He's so single minded, never goes out, no social life or cocktail parties. He's thoroughly absorbed, goes home and thinks about the legislation.[20]

Forcing or browbeating was never Mills' style. He was, in all respects, a gentleman and ran his committee in a genuinely nonpartisan manner, treating Republican members with the same elaborate courtesy with which he treated the Democrats.

More than anything else, what really contributed to Mills' power was his ability to compromise, to seek a consensus, to know what position or modification here and there would garner the widest base of support in committee. Mills himself has agreed about the process through which he put his committee when it considered legislation:

> Oh yes, there is no place for just the idealist because we're dealing with very practical matters. We're living in a very practical world. We have to find practical answers.[21]

Wilbur Mills' own practicality, however, seemed to desert him in 1971, a time when his power was never greater or more openly recognized. It is almost as if Mills started reading his press clippings in 1971 for the first time. *Reader's Digest* published a feature story on him captioned "The Most Powerful Man in Congress." Hugh Sidey, in his weekly column on the presidency for *Life* magazine, described Mills in awesome terms:

If the White House were to rate the independent powers of
this world with which it must deal according to their formidability,
Arkansas Congressman Wilbur Mills probably could rank just
below Germany and above France. Or that at least is the way they
tell it.[22]

Sidey then went on to report the obsequiousness with which the
Nixon White House was then treating Mills in an effort to gain
his support for its revenue-sharing proposals.

Indeed, at one point in 1971, Mills was running his own little
"state department" and conducting his own foreign policy. In
1970, at the request of the Nixon Administration, Mills, himself
a free trader, introduced a textile quota bill. Nixon had told Mills
that the introduction of such a bill by him could be used effectively
by the United States as a lever for negotiations with the Japanese.
Nixon proved wrong and the Japanese did not budge. In January
of 1971, Mills introduced another such bill, again as a means of
putting pressure on the Japanese. Nixon Administration officials
were still unable to make any headway with the Japanese in their
official talks.

The Nixon White House then decided to try a more elaborate
tactic. It prepared for Mills a strongly worded speech calling for
strict protectionist measures. Mills called Japanese Ambassador
Nobuhiko Ushiba to his office in early February and handed him a
draft of the speech, suggesting that he intended to deliver it. Ushiba
was shaken that a free trader like Mills would give such a speech.
He asked Mills not to deliver the speech until he could consult
with the Japanese government.[23]

Shortly after this meeting, in a conference with a delegation of
Japanese textile manufacturers, Mills went even further and suggested that, instead of a negotiating meeting between the two
countries regarding textile imports, the textile manufacturers should
unilaterally declare a limitation on their shipments to the United
States. Washington attorney Michael Daniels, at Mills' suggestion,
traveled to Tokyo to assist in drafting the unilateral statement by
the Japanese textile manufacturers. The document was then brought
to Mills for his review. He suggested a few changes, the Japanese
agreed, and after this agreement was announced in Tokyo, Mills

publicly gave his approval. Both Secretary of State William Rogers and National Security Adviser Henry Kissinger urged Nixon to accept the unilateral declaration of the Japanese. Jealous White House aides, who had been involved in direct negotiations with the Japanese government, and whose feathers were ruffled by Mills' personal diplomacy, eventually persuaded Nixon to reject the unilateral Japanese declaration.

Given his involvement in such high-level international politics, perhaps it is not surprising that Mills-for-President rumors first began floating early in 1971. Charles Ward, an Arkansas businessman, had conceived the idea of drafting Mills for president and had spoken with Mills about it to insure that he had no objection. According to Ward, Mills did not object, and in July, 1971, Ward opened a campaign office in Washington, D.C., for the "Draft Mills for President Committee" with Ward as the Chairman.[24] Mills has since disclaimed any responsibility for the Draft Mills Committee, but there are reasons to take these denials with a grain of salt. For one thing, Mills never made these denials under oath. Moreover, given Mills' presumed authority in his home state of Arkansas, it is difficult to believe Mills had no role in the Arkansas Legislature's amendment of their election laws early in 1971 to permit Mills to run simultaneously for re-election to Congress and for president or vice-president. Legislation similar to this was passed by Texas in 1960 which permitted Lyndon Johnson to run both for re-election to the U.S. Senate as well as for vice-president.

In public, however, Mills was playing it straight. His initial reactions ranged from flattered disbelief upon receiving unsolicited campaign checks in his congressional office to accusations that the rumors were being floated by the Nixon Administration in order to discredit him.

Nevertheless, like all "draft" candidates, Mills refused to take himself out of the picture, conjuring up visions of extreme circumstances which would cause him to make the ultimate sacrifice:

> I'll tell you, I have absolutely no interest, but if that fellow downtown doesn't straighten up after a while, and it would take me running on the ticket to beat him and there would be no other way, then I think I'd do it.[25]

Mills continued to play the game. He told *U.S. News and World Report* that he was not a candidate but that "if the convention did what I consider the impossible, I would accept the nomination. . . . Should I be nominated, I'd be as active a candidate as the Democratic party ever had."[26] It was hardly a retiring type of non-candidacy.

The proposed Draft Mills strategy was scarcely remarkable. It involved such items as revealing identities of well-known supporters of Mills, a few at a time, in order to create the illusion of momentum; establishing good relations with local politicians, state legislators, and governors through Mills' various travels; portraying Mills as "the most qualified man to run the government" because of his expertise from the Ways and Means Committee on taxes, Social Security, and foreign trade; maintaining good relations with both labor and business; and, inevitably, hoping for a deadlock at the convention which would allow him to be presented as the ideal "compromise" candidate.[27]

Candidates for president, coy or otherwise, cannot get by without money, and Wilbur Mills was no exception. He had been a friend of dairy interests for years and, like Hubert Humphrey, was quite prominent in the dairy lobby's efforts on Capitol Hill in the spring of 1971 to reverse the Nixon Administration's initial negative decision on raising milk-price supports.

In February of 1971, Mills had Speaker of the House Carl Albert summon Agriculture Department officials and White House liaison personnel to the speaker's office to meet with him, Mills, and the dairy producers to let the administration know just where the two most powerful men in Congress—the Speaker of the House and chairman of the Ways and Means Committee—stood on milk.[28]

Dr. George Mehren, then a consultant to AMPI, talked with Mills repeatedly during the late winter of 1971, met with him on March 16, 1971, and worked directly with Mills to disseminate data to other Congressmen on the milk price support decision, as well as to assess with Mills the chances of passing the legislation. Harold Nelson, then AMPI General Manager, also worked directly with Mills in orchestrating congressional support for legislation.[29]

The Nixon Administration certainly felt Mills' pressure on this issue. John Connally testified about a number of conversations Mills had with him between March 12 and March 25, 1971, about the price support decision.[30] Likewise, Clark McGregor, the president's assistant on Congressional affairs, was lobbied by Mills on six separate occasions in February and March of 1971 to urge the president to raise the milk support price.[31] Mills also contacted the director of the Office of Management and Budget, George Schultz, on March 4, 1971, the purpose being, says Schultz, "to push for a prompt decision" on the price-support increase.[32]

Such loyalty by Mills did not go unrewarded when he started thinking about the 1972 Democratic nomination for president. Mills received legal contributions from the political action committees formed by the dairy lobby; individual contributions from members, employees, and officers of the dairy cooperatives; and illegal contributions from corporate assets of dairy cooperatives. The total, illegal and legal, amounted to almost $200,000.[33]

Mills' legal dairy contributions from TAPE, CTAPE, ADEPT and SPACE totalled $55,600, $26,500 of which came from AMPI's political action committees, TAPE and CTAPE, the bulk of it being received in August, 1972. The rest came from ADEPT, Mid-America Dairymen's political trust fund ($16,600 in June, July, and August of 1972), and SPACE, Dairymen, Inc.'s trust fund ($12,500 in May, June, and August, 1972).[34] The pattern was relatively clear. All of the legal contributions from the political trust funds of the dairy lobby were made to Mills during his *announced* candidacy stage. The illegal contributions came largely during the undeclared stage of Mills' candidacy, the so-called Draft Mills movement over which Mills claimed to have had no control.

Other legal dairy contributions to Mills, albeit of more dubious origins, came from individual employees, officers, and directors of AMPI. The man responsible was David Parr, who, in late 1971 and early 1972, raised $40,000 for Mills through personal solicitations. Parr claimed that no pressure was used, although he admitted he had heard of some "arm twisting."[35] Parr was in charge of personnel solicitation, however, and if there was any muscle involved, Parr can hardly disclaim responsibility. After extracting, from time to time, these contributions from the AMPI personnel,

Parr would then periodically deliver to Charles Ward, chairman of the Draft Mills Committee, $4,000 to $5,000 in checks. Parr also had AMPI deliver checks to Gene Goss (Mills' administrative assistant) at Mills' congressional office.[36]

Parr also was not above forcing AMPI attorneys to contribute for Mills as well. Stuart Russell gave $1,000 to Mills, as did attorney Frank Masters. Russell was reimbursed, as per his usual arrangement with AMPI. Masters, however, denied any reimbursement by AMPI.[37]

Parr's protestations of ignorance regarding strong-arm tactics is further undermined by the fact that he also attempted in the early part of 1972 to initiate a checkoff system for contributing to Mills' presidential campaign. Initially, the checkoff system was limited to employees in AMPI's southern region. It involved sixty-five employees, who were programmed by Parr to contribute $25 a month each. His boss, Harold Nelson, knew that Parr was putting pressure on key AMPI employees and that he was personally approaching them. "Mr. Parr is a pretty forceful character," said Nelson.[38] Fortunately for AMPI's employees, Parr's new boss was even more forceful. When Dr. George Mehren, as AMPI's new general manager in January, 1972, learned of Parr's grandiose schemes for promoting Mills (through complaints from coerced employees), he stopped Parr dead in his tracks, and the checkoff was never implemented.[39]

Most politicians, for obvious reasons, claim to be largely unaware of the sources of their various campaign contributions. Mills was no exception. In AMPI's case, however, he learned about it. With particular reference to the $25,000 contribution to Mills' campaign on June 13, 1972, from CTAPE, George Mehren, at that time general manager of AMPI, testified that Mills had known of it and had thanked AMPI for it. Mehren also conceded that the campaign contribution was, in part, a payoff for Mills' help with the 1971 milk price support decision. "I think, in complete honesty," said Mehren, "it [was] an incident in a long history of understanding, awareness, accessibility and support."[40]

AMPI's illegal contributions to Mills were far more important than the legal contributions which were to follow. "Up front" money before their presidential campaigns catch the public imagination is the most difficult for candidates to find. AMPI filled that need for Mills with men, money, and material.

In any political campaign, willing and able workers are as valuable as campaign funds. Betty Clement, Joe Johnson, and Terry Shay were three of those willing workers for Wilbur Mills. And therein lies a tale for which it is difficult for Mills to escape personal responsibility.

Dave Parr and Harold Nelson were directly responsible for assigning other AMPI employees to work in the Mills campaign: Joe Johnson, who served directly as a campaign aide to Mills and traveled with him; Betty Clement, who worked in the office of the Draft Mills for President Committee, and Terry Shay, who was an advance man. All of them worked full time for the Mills campaign. The combined monthly salary of the three individuals was $2,500 and while Mills' administrative assistant, Gene Goss, denied all knowledge of who was paying them, Dave Parr is certain that "they knew it. We did not try to hide it." Goss admitted that in a trip to Arkansas in February, 1972 to meet with Mills, he recommended that Joe Johnson be put on Mills' congressional payroll. If Goss were suggesting to Mills what payroll Johnson ought to be on in the future, it is reasonable to assume he had some idea of whose payroll he had been on in the past, particularly since Johnson had been working full time in the Mills campaign for over eight months.

Betty Clement went to work for AMPI in 1970 as a secretary at its Little Rock office. She continued to work for AMPI (or at least thought she did) until February 1, 1972. What she did not know was that AMPI took her off its payroll on June 30, 1971, and that her salary was thereafter paid by Warren K. Bass, a certified public accountant in Little Rock who was reimbursed for these expenses by AMPI. Bass and Mills had formed a nonprofit association called the Arkansas Voter Registration Association (which was succeeded by NVRA, the National Voter Registration Association), the ostensible purpose of which was to educate the public on voter registration and to do "research."[41]

NVRA was used by AMPI and Mills to hide the illegal support being given to his campaign. Clement, for example, (whose salary was still being paid by AMPI using Bass as a conduit) was assigned in July, 1971 to work on the Draft Mills Campaign. She worked for two months at Mills' headquarters in Little Rock, and she next went to work to prepare for an allegedly nonpartisan rally in Ames, Iowa. In November, 1971, Clement was sent to Washington, D.C., where she worked on Mills' campaign until February 1, 1972. After that, Mills put her on his own Congressional payroll once George Mehren terminated her employment with AMPI.[42]

The whole NVRA scheme to mask AMPI's contribution of campaign workers to the Mills campaign began in 1969 with discussions between Dave Parr and Warren Bass. Bass agreed that NVRA would furnish AMPI with voter registration information, educate key AMPI personnel and AMPI members and "supply Mr. Parr with any other information he might request," in exchange for which AMPI would pay all of the salary expenses of NVRA, which eventually totaled $1,750 a month. The employees then working for NVRA, like Clement, would subsequently work on the Mills campaign.

Clement, of course, simply did what she was told. She was not paid much, only $750 a month, but that was $750 a month which Mills saved. Clement was assigned by "Dave Parr and the rest of them" to take charge of the Draft Mills headquarters office in Little Rock, where she answered the telephone, paid bills, and kept the books. She took time out to attend the AMPI annual meeting in Chicago in September of 1971 and then returned to AMPI's Little Rock office for the rest of September and October, 1971.[43]

When she went to Washington, Clement got a free apartment and furniture as well. She signed a lease for apartment 611 at the Watergate Towers, 907 Sixth Street, S.W., Washington, D.C., for a twelve-month period beginning December 1, 1971. The monthly rent was $298. She also leased furniture for the apartment at $77.80 a month. AMPI paid these expenses for her apartment through January, 1972.[44]

Before Joe Johnson went to work on the Mills campaign, he had been Director of Field Services for AMPI's Northern Texas division and had been making $25,000 a year. Mills was an old friend of the Johnson family, had known Johnson's father, and had offered to appoint Johnson to West Point in 1947.[45]

In July, 1971, Dave Parr asked Johnson to go to Little Rock to plan an appreciation dinner for Mills. Johnson did so and spent ten to fifteen days in Little Rock working on the project. Terry Shay helped Johnson on this project also, and both their expenses and salaries were paid by AMPI. Thereafter, Johnson continued to work on the Mills campaign, sometimes as much as eighteen hours a day. Johnson traveled on campaign business with Mills to Chicago, Miami, and California, AMPI picking up all his expenses. In November, 1971, Johnson also moved to Washington (at the same time as Shay and Betty Clement) and received a free apartment and furniture in the same building as Clement. Johnson was also active in organizing Mills' campaign for the New Hampshire Presidential Primary early in 1972, and AMPI reimbursed him $2,850 for his New Hampshire campaign expenses.[46]

Immediately before the New Hampshire primary, Johnson resigned from AMPI and became a paid staff member of Mills' congressional staff. Thereafter, Johnson continued to participate in campaign work outside Washington, D.C., and at the end of March 1972, he was taken off Mills' payroll because of advice from the clerk of the House forbidding campaign activity by a congressional employee. Johnson then signed on as chairman of the official Mills for President Committee.

Other AMPI representatives also devoted time to the Mills campaign while on AMPI's payroll. For example, former Arkansas State Senator Charles George and his wife had an expenses paid excursion to New Hampshire, furnished by AMPI, the purpose of which was to boost the Mills candidacy. Another AMPI employee with a background in radio and television promotion was sent by AMPI on an expenses paid trip to New Hampshire. In all, there were as many as four or five other AMPI employees working off and on in the Mills campaign.[47]

AMPI furnished illegal cash contributions to Mills, all of which followed the usual AMPI pattern and were laundered through conduits. On August 17, 1971, Harold Nelson told Bob Lilly to deliver $5,000 in cash to Parr. Lilly promptly went and personally borrowed $10,000 at the Citizens National Bank of Austin. Nelson also told Lilly that he could repay the loan by getting the money from various AMPI attorneys on retainer, "and the attorney in turn would bill AMPI double," presumably to cover the attorney's tax liability on the money received from AMPI.[48] According to Parr, the $5,000 was for a Wilbur Mills appreciation dinner.

After borrowing the $10,000 at Citizens National Bank, Lilly drove out to the Little Rock Airport, where he delivered $5,000 to Parr's secretary, Norma Kirk, who went back to Parr's office and placed the money in a walk-in vault. Parr then told his AMPI subordinate, Tom Townsend, to take the money to Washington, D.C.[49] Townsend flew there, met with Gene Goss, administrative assistant to Wilbur Mills, and gave him the sealed envelope, telling him it was from Parr.

Goss, for his part, claimed he had no recollection of receiving an envelope from Parr via Townsend. The only thing Goss would admit to receiving was the many checks from individual dairy farmers and AMPI employees, whose contributions Parr had coerced (since the contributions were by check, Goss could not very well profess a lack of recollection of them anyway).

Still another allegedly illegal contribution to the Mills campaign —also for $5,000—was arranged by Dave Parr. Parr had called Texas lawyer Jake Jacobsen, who had long been on a retainer arrangement with AMPI. Parr told Jacobsen that he wanted a $5,000 contribution to Mills' presidential campaign from Jacobsen. Thereafter, Jacobsen's law partner, Joe Long, made out two checks on November 10, 1971, one for $2,750 and the other for $2,250, cashed them, put the money in an envelope and drove out to the airport to meet with Jacobsen, Dave Parr, and Tom Townsend. At the airport coffee shop, Long handed the envelope to Jacobsen. "Here's the $5,000 for Wilbur that you wanted. . . ." Jacobsen took the envelope and passed it over to Parr, "Here's the $5,000 for Mr. Mills." Parr put the envelope in his pocket. Jacobsen subsequently denied being reimbursed by AMPI for this contribution although it was his practice on other occasions.[50]

AMPI's only other illegal cash contribution to the Mills campaign involved $9,291.53, which AMPI paid for the printing of

110,000 bumper strips ordered by Dave Parr in the summer of 1971 and carrying the legend "Mills for President." The invoice was sent directly to "Associated Milk Producers Assoc.," to the attention of Dave Parr.[51] One hundred ten thousand bumper strips were a big order for a noncandidate like Mills, but Parr had a good reason for placing such an order. He and Wilbur Mills had big plans for that fall, plans which involved the heretofore non-partisan Iowa Institute of Cooperation.

The Iowa Institute of Cooperation is an association of farmer cooperatives in Iowa. Its objectives include educational training, legislative lobbying, and state-wide public relations work to maintain a favorable climate in which cooperatives can thrive.[52] For many years the Iowa Institute of Cooperation had the governor of Iowa proclaim a co-op month, and 1971 was no different— early that year, the month of October was so designated. In the past, the Iowa Institute of Cooperation had never held a rally during co-op month and as of September, no rally was planned for co-op month in 1971 either. According to Gerald R. Pepper, executive director of the Iowa Institute of Cooperation, "there was never an intent at that point for there to be a program. . . . There was no rally planned. . . ."[53] Enter Wilbur Mills. On Labor Day, Mills called Pepper:

> Mr. Pepper, we have powerful problems in agriculture. . . . I wonder if you would do me a personal favor [and] rent the University of Iowa Football Stadium and fill it with farm people and give me an opportunity to come out and meet with [them].[54]

Pepper was to call Mills back the following night at Mills' apartment. The more Pepper thought about it, the more he liked the idea. After all, it was co-op month, and Mills' suggestion was, in Pepper's opinion, "a tremendous opportunity to focus attention on this program." Pepper called Mills back and told him he agreed with the idea. Mills replied that he would have someone get in touch with him.[55]

Mills moved quickly. Joe Johnson of Mills' campaign staff called Pepper, identified himself as a representative of AMPI, and told Pepper that he would have no worry about money, that AMPI would take care of all financial arrangements. Johnson called Pepper later that same day, Tuesday, and asked him if he could come to Washington on Friday. Pepper replied that he could not do so because he had a co-op board meeting on that date.

Minor matters like that were no problem for Johnson, who grandly offered to transport the entire board to Washington so they could have their meeting on the airplane. Johnson subsequently arranged for two private jets to transport the ten members of the board to Washington. Johnson was on the plane with Pepper and advised him that in Washington he would be expected to invite Wilbur Mills to address the rally.

Upon landing in Washington, the entire board was taken to the ornate high-ceilinged Ways and Means Committee room, where a large group was meeting. Dave Parr, who was presiding, made a few introductory comments and, when Mills arrived, Parr introduced him to Pepper. Pepper dutifully proffered the requested invitation and Mills accepted. Afterwards, the co-op board conducted a brief meeting in the Ways and Means hearing room, suitably awed by their proximity to power.[56]

AMPI swung into high gear to promote the Mills rally. Johnson arrived in Ames, Iowa with a large contingent of AMPI staff members, including Forest Wisdom, John Holmes, Tom Townsend, and, of course, Terry Shay and Betty Clement. Installing a bank of phones, they used the local Holiday Inn as their central headquarters. Throughout the month of September, AMPI had a contingent of at least six employees in Ames working on the Mills rally with their complement sometimes swollen to fifteen or twenty.[57]

AMPI kept a tight rein on the finances of the Mills rally. Johnson established a bank account in a local Ames bank and deposited the necessary money to cover the cost of the rally. While Johnson had the power to draw on the bank account, the officials of the Iowa Institute of Cooperation did not. Pepper even turned over to Johnson a $15,000 check he had received from Mid-American Dairymen. Whenever Pepper had bills, he simply delivered them to Johnson for payment.[58]

In total, AMPI deposited over $38,000 in the account opened by Johnson in Ames, Iowa. In addition, AMPI directly paid another $6,132, including $3,751 for buses chartered to transport farmers to and from the rally. In the meantime, Pepper was freely traveling around the state by aircraft, while Johnson (and AMPI) picked up the tab. When Pepper needed air transportation, he would merely call Johnson, who would identify the aircraft for Pepper and tell him when and where it would be at the Ames airport.[59]

While it is difficult to calculate with precision the total amount of money spent by AMPI on the Mills rally, it is apparent that, together with Mid-America Dairymen, Inc., direct payments from corporate assets totaled $44,132. In addition to that, AMPI paid the salaries for an average of six employees for the entire month of September. Add to this the expense of the two jet aircraft in which Johnson grandiosely arranged to bring Pepper's board to Washington, and the amount easily exceeds $50,000.[60]

The question is, of course, whether it was really a rally for Mills and his presidential candidacy. All the available evidence points to the inescapable conclusion that it was. After all, Mills suggested the very idea of the rally and then solicited his own invitation to do it. Other evidence comes from Dave Parr, who conceded that the rally was designed to give more prominence to Mills and his undeclared campaign.

Mills' campaign for the presidential nomination eventually fizzled in the snow of New Hampshire. While he kept his campaign alive right up to the convention, the imagined deadlock he was waiting for never occurred. Less than two years later, his first and last foray into national electoral politics was to return to haunt him.

Nineteen seventy-four was not a good year for Wilbur Mills. It started out bad and got worse. In January, Senator Sam Ervin of the Senate Watergate Committee sent Mills a letter:

> Dear Congressman Mills:
> As you know the Senate Select Committee on Presidential Campaign Activities is investigating various allegations concerning the employment of corporate funds by Associated Milk Producers, Inc., and others for the benefit of various political candidates in the 1972 presidential campaign and election. In this regard, we are interested in obtaining certain information and materials from you, of course, at your convenience.
> While we know that your schedule is extremely tight, the Com-

mittee would appreciate it if you would consent to meet with members of our staff. Members of our staff will be in contact with members of your staff in the very near future.

Thank you in advance for your cooperation.[61]

Mills was obviously concerned because, following the time-honored tactic of cornered political animals, he froze. Like Hubert Humphrey, Mills ignored Ervin's request. When he received no response, Senator Ervin tried again on February 7, 1974:

> This letter is intended as a follow-up to my letter to you on January 24, 1974, and is designed to provide certain specifics regarding the inquiries the committee wishes to make of you.
>
> It . . . appears from evidence gathered by the committee that Associated Milk Producers, Inc. may have made corporate contributions to your campaign, and may have paid for certain services rendered your campaign by several individuals.
>
> We have absolutely no knowledge that you were contemporaneously aware of these circumstances. Nevertheless, . . . we feel it necessary to speak with you respecting these matters. The committee would also like to examine certain records in your files and our staff will promptly contact your staff to specify the records we would like to see.[62]

Again, no response. Mills then hired a lawyer and, on March 18, had him tell Senate Watergate Committee lawyers that Mills would be willing to meet with a senator from the committee as soon as the House had completed action on certain legislation. His attorney assured the committee lawyers that Mills would notify Senator Ervin by letter when he was ready to meet. But Mills never wrote to Ervin. Like Hubert Humphrey, Mills never talked to Ervin or to any other senator on the committee regarding his presidential campaign.[63] Even when asked to furnish his logs and calendars for the February and March period to the Senate Watergate Committee, Mills refused for the rather improbable reason, according to his attorney, that "such records have not been maintained."[64]

Mills' old campaign manager, Joe Johnson, continued the Mills stonewall tactics. On April 2, 1974, Johnson was called to testify before the Senate Watergate Committee, took the Fifth Amendment and refused to answer any questions. Senator Herman Talmadge, chairman of the committee session that day, sustained Johnson's refusal to testify on the basis of his self-incrimination privilege and made no effort to compel him to testify by granting limited-use immunity.[65]

While Mills could not find time to talk to the Senate Watergate Committee, he did find time to appear on a Washington talk show for the Public Broadcasting System to answer questions regarding his presidential campaign financing. In response to some soft questions, Mills admitted receiving substantial contributions from the dairy lobby but explained it away by claiming that "the milk people have always been friendly with me" because "I've been aware of their problems." In the next breath, and without an intervening question, Mills falsely went on to deny that he had ever played any role in pressuring the Nixon Administration to increase milk price supports.

Mills had to be deeply worried about the illegal contributions. He even volunteered that "there were apparently some moneys that apparently . . . were given long before I became a candidate," but denied that he had had any knowledge of this. "I was unaware," he claimed, 'of what was going on in the so-called Draft Mills period."[66]

No one was ever able to prove this false because, as the Senate Watergate Committee concluded:

> A full development of the circumstances [of illegal contributions to the Mills campaign] has been precluded because of Johnson's invocation of the Fifth Amendment when subsequently called to testify under oath, and by the failure of Mr. Mills to accede to the committee's request for an interview.[67]

The pressure on Mills during this time must have been intense. Wielding his vast power as chairman of Ways and Means had been done quietly and smoothly in the back halls and committee rooms of Congress. There Mills was in charge and he knew what he was doing. As a presidential candidate, Mills had been out of his element. Now, as the target of a campaign financing investigation by the Senate Watergate Committee, his only defense was the same as that of Hubert Humphrey and Richard Nixon—he did not know what his aides were doing in his behalf.

The pressure proved too great. It first surfaced publicly early on a Monday morning in October 1974. Mills had been out with a group of two men and three women. One of the women, Annabella Battistella, was a neighbor and close friend of Mills and his wife, Polly, who wasn't with Mills that night. She was laid up at home with a broken foot so Mills left her behind and later claimed that they were throwing a bon voyage party for Mrs. Battistella's cousin, Gloria Sanchez, who was returning to Argentina. A harmless party, perhaps, and with a little luck nothing would have come of it. Just one more rumor added to the many floating around Washington. Unfortunately for Mills, the party lasted too long. At 2:00 A.M. that Monday morning, the Mills party was returning home in his blue Lincoln Continental driven by Albert Gapacini. Caught in the revelry of the moment, however, Mr. Gapacini had neglected to turn on the headlights. The absence of headlights drew the attention of two District of Columbia park police officers as the car headed toward the Jefferson Memorial. The police gave chase and managed to stop the car in the vicinity of the Tidal Basin.[68]

The Mills party tumbled out of the car in an advanced state of inebriation. Mills, however, was not only drunk but had apparently also been engaged in some sort of struggle with Mrs. Battistella: his face was scratched, and his nose was bleeding, and Mrs. Battistella was sporting two black eyes. Upon emerging from the car, Mrs. Battistella ran for the Potomac and promptly plunged in. Whether she was seeking to escape the police by swimming to the opposite shore or was merely too drunk to know what she was doing, the police dutifully plunged in after her and dragged her back to shore.[69]

The District of Columbia police report of the incident charitably omitted all names. Two days after the incident, Gene Goss, Mills' administrative assistant, was still stonewalling the press and insisting that, while the dark blue Continental might have been Mills' car, Mills was snug at home safe in bed at the time of the accident. Eventually, however, the District of Columbia park police finally admitted that Mills had, in fact, been in the car at the scene of the accident.[70]

Goss explained to the press that he must have "misunderstood" what the congressman had told him earlier. Goss then went on to say that Mrs. Battistella had this cousin—Gloria Sanchez—who was going to return home to Argentina. Congressman Mills, because he and his wife Polly were such good close friends of Mrs. Battistella and her husband Eduardo, wanted to throw a bon voyage party for Gloria Sanchez. But Mrs. Mills could not come to the bon voyage party because she had a broken foot. So she had to stay home. For some reason, Mrs. Battistella's husband Eduardo could not make it either. So, Mills and the two women plus Al Gapacini and a third woman headed out for a Caribbean nightclub called the Junkanoo. After a good time at the Junkanoo, they went to another "public place." After having a "few refreshments," Mrs. Battistella became ill and the party had to call it a night. But when they were driving home in the Mills Continental, Battistella suddenly tried to leave the car. The congressman had to try and stop her. Mrs. Battistella obviously wasn't very appreciative of the congressman's solicitous help because she promptly gave him a sharp elbow in the face, breaking his glasses and inflicting cuts on his face. In other words, the congressman got hurt just trying to help, a fate similar to other good Samaritans.[71]

The embarrassing part of the story for Mills, however, was that Mrs. Battistella did not always use her married name. Others knew her as "Fanne Foxe, the Argentine Fire Cracker," who worked as a stripper at the Silver Slipper, a well-known Washington night spot.[72]

Faced with an unexpectedly strong challenge for re-election in November from a young and aggressive Republican candidate, Judy Petty, Mills was re-elected by the relatively narrow margin for him of 59 percent to 41 percent. By this time, the pressures of Watergate and his own milk fund scandal had already taken their toll. Mills literally had lost control of his actions. In December 1974, he made a little side trip to Boston to catch Mrs. Battistella's act at the Pilgrim Theatre, a rundown burlesque house in Boston's "combat zone," famed for its pornographic movies and live striptease shows. Mills was, to say the least, indiscreet that night. Fanne Foxe had just finished her act, clad only in a red wig and black g-string. After leaving the stage for a moment, she re-entered after donning a diaphanous pink gown. She announced to the audience that there was someone she would like them to meet. "Mr. Mills, Mr. Mills, where are you?" she called. Mills, who had watched Fanne's performance from backstage and whose purpose in following Fanne to Boston was, he claimed, to lay to rest rumors that he was having an affair with her, bounded onto the stage, shook Fanne's hand, gave her a brief hug on the shoulder and left.[73]

And with him went his career. The "Watergate Class" of freshman Democratic congressmen had just entered the House of Representatives, they were in a rebellious mood, and what they regarded as a senile old fool heading the most important committee in Congress—the House Ways and Means Committee—was the last thing they were prepared to tolerate. After his stage debut in Boston, the House Democratic Caucus stripped Mills of most of his power as Ways and Means Committee chairman. The authority of the Democratic members of the committee to make all other committee assignments to Democratic House members was taken away. The Democratic Caucus also voted to increase Mills' committee to thirty-seven members, twenty-five of whom would be Democrats, thereby diluting the conservative Democratic and Republican coalition on the committee, which Mills had forged over the years.[74]

After these humiliations, Mills checked himself into Bethesda Naval Hospital, declared himself an alcoholic, and retired from Congress at the end of his term. In a real sense, the dairy lobby had succeeded, however inadvertently, in ending the career of the most powerful man in Congress. In many ways—and certainly to Mills—it was a greater individual tragedy than even that of Richard Nixon. Unfortunately for Mills, the final demise of his career occurred in such a comic fashion that no one stopped to ask why such a respected figure—described by one long-time Republican colleague on Ways and Means as "the brightest man I ever met"—had come to such a sad end. The answer, of course, is that his reach exceeded his grasp—he convinced himself that he was the perfect compromise candidate for the Democratic presidential nomination in 1972. Many people in Washington probably had similar daydreams. But they had no dairy lobby willing to bankroll their dreams. Mills did, and the final, sad outcome was Fanne Fox and the Washington Tidal Basin.

No one stopped to ponder all this in late 1974 because the news was being monopolized by Richard Nixon's resignation and its aftermath. The dairy lobby's responsibility in that drama was far greater than it had been for Mills. Not only did the dairy lobby play a major role in Richard Nixon's impeachment proceedings, but its actions also led directly to the bribery trial of Nixon's treasury secretary, John Connally.

5

The Republicans: Buying "Access" to the Nixon Administration

Having been a heavy contributor to Democratic causes for many years, the dairy lobby found itself in an awkward position when the Nixon Administration came to power in 1969. It had backed a loser, Hubert Humphrey, but still needed powerful friends in government. The dairy lobby was to prove, however, that you did not need to contribute to a president's campaign finances before his election in order to buy his friendship. Substantial contributions after the election would do quite nicely.

In pursuit of such friendship, AMPI's Harold Nelson and Dave Parr talked about "the possibility of making some contributions to the Republican party" in 1969 with one of their lawyers, Jake Jacobsen. To Harold Nelson, the problem was one of gaining "access."[1]

To win a "more sympathetic ear in the Republican administration, since [the dairy lobby] had supported Senator Humphrey,"[2] Jake Jacobsen proposed to use his Washington law partner, Milton Semer, an unlikely man to solve this particular problem, since he and Jacobsen had both worked in the Johnson White House and were lifelong Democrats. Moreover, Semer was a major fund-raiser for Maine's Senator Edmund Muskie, who had been Humphrey's running mate in 1968.

Semer met Jacobsen, Parr and Nelson early in 1969, when the dairy lobby was still undecided on the best way to approach the new administration. Semer viewed his task as guiding his new clients through the government bureaucracy by "finding out how the White House would be organized to handle special interest groups, . . . who would handle their problems . . . and coordinating their lobbying efforts in Washington."[3]

Semer's connection with the Nixon Administration was tenuous. When Semer had been General Counsel at HUD during the Johnson Administration, John Mitchell, a well known municipal bond lawyer, had served on an advisory committee to the General Counsel's office.[4] Therefore, after one of Semer's clients asked him how he might make a substantial contribution to the Nixon campaign, Semer called Mitchell, then the Nixon Campaign Chairman, and was referred to Maurice Stans, who shuffled Semer over to Jack Gleason, at that time an assistant to Stans in the fund-raising effort for the 1968 campaign.[5] When the dairy lobby came to see him after Nixon's election, Semer decided to again call Jack Gleason.

The first contact between Semer and Gleason took place on March 25, 1969, four days after the initial Nelson/Parr/Jacobsen/Semer meeting. Semer's purpose was to familiarize Gleason with the dairy lobby's interest in government farm policies and to advise Gleason of the fund-raising efforts of the dairy lobby's new political action committees and their plan to contribute substantial money to congressional and presidential candidates. Gleason expressed a keen interest in the political funds, and suggested that a meeting be arranged with Herb Kalmbach, who, Gleason candidly admitted, was not a member of the government but "might have insights" that Gleason lacked.[6]

In 1969, California attorney Herbert Kalmbach was just starting to bathe in the blissful notoriety of being the president's personal lawyer, and such publicity, when meshed with Gleason's recommendations, convinced Semer that "Kalmbach was emerging . . . as an important and influential advisor to the president and the White House, along the lines that people such as Clark Clifford . . . and others had performed for the Democrats."[7]

Moreover, Kalmbach was the trustee of an important sum of cash left over from Nixon's 1968 campaign. Kalmbach understood that, although he was the actual trustee of the fund, it was H. R. Haldeman who was in full control of the money.[8] Kalmbach's job was to keep the funds under control and to report to Haldeman for all expenditures and all receipts. Haldeman told him to keep the cash as cash, and the checking account funds in the checking account.

Haldeman further made it clear to Kalmbach that if he was offered cash or could obtain cash, rather than checks, he should do so because the Nixon team was depleting the cash funds at a rapid rate. Kalmbach was so successful at following Haldeman's instructions that by 1972, the fund amounted to almost $1 million, fully one-quarter of which was in cash.

On the morning of April 3, 1969, Semer and Kalmbach met in Washington, D.C., for the first time. Semer explained to Kalmbach, as he had to Gleason, the nature of the dairy lobby, its interest in government farm policies, and the fund raising potential of the political action committees of the three major milk co-ops, AMPI, DI and Mid-Am. Kalmbach was very interested in the potential for contributions from the dairy lobby, after Semer's assertion that Nixon could expect a potential "one million dollars a year."

During the spring of 1969, Semer and Kalmbach had several more meetings at which the specific objectives of the dairy lobby were spelled out: it wanted milk price supports raised to a high level, 90 percent of parity, and it wanted the name of a White House aide who could serve as its contact point or liaison with the Nixon Administration. Although not as important, Semer also indicated his client's desire both for a personal meeting with the president, and an appearance and speech by Nixon at AMPI's 1970 convention in Kansas City.

After these initial meetings with Kalmbach in April and May, Semer went to Dallas on July 9, 1969, to meet with Nelson, Parr, and Jacobsen prior to his seeing Kalmbach on the next day in California. The purpose of the Dallas meeting was to make an assessment of the situation both in terms of money and access. Not much progress had been made in getting information from Gleason, Kalmbach, or others at the White House, and hence the dairy lobby was still unaware of how the Nixon Administration would be organized to handle special interest groups. Nevertheless, Semer was authorized to promise Kalmbach a $100,000 cash contribution.

Kalmbach, by the time he met on July 10, 1969, with Semer, was more than ready to do business. He had already confirmed with John Mitchell that Mitchell knew Semer. Further, Haldeman had approved additional meetings to discuss arrangements for receiving contributions. Armed with the $100,000 authorization from his clients, Semer was ready too, and a deal was quickly made for a cash contribution to Kalmbach in that amount.

Following Semer's meeting with Kalmbach in California, Harold Nelson told Bob Lilly to pick up $100,000 in cash from TAPE and give it to Semer, who would deliver it to Kalmbach "to get the favorable attention of the Republicans" as an "offset" to AMPI pro-Democratic activity in the 1968 elections.[9]

Lilly contacted the Citizens National Bank in Austin, Texas, a depository for TAPE funds, and worked out the details of this rather sizeable transaction with the president of the bank, Marvin Stetler, who had been briefed earlier about the large withdrawal by Jake Jacobsen, who was also the bank's chairman of the board. Stetler told Lilly that it would take several days to accumulate that much money in hundred dollar bills from several banks, in order to avoid the curiosity of banking authorities in the federal government. Once gathered, the money was debited to the TAPE account and readied for Lilly to pick up on August 1. Lilly then called Semer in Washington, D.C., to arrange for a meeting in Dallas. On August 1, Lilly went to Austin, visited Stetler's office, watched him count out $100,000 in hundred dollar bills, scooped the money into his briefcase, and headed for the airport to fly back to Dallas. Once there, he took a short taxi ride to the Executive Inn and went to Semer's room. Semer took the briefcase without counting the money or giving Lilly a receipt and promptly hopped on a plane to California.[10]

Upon his arrival in Newport Beach, Semer delivered the cash to Kalmbach, who commingled it with other Nixon funds in a safety deposit box at a nearby bank. Once the money was safely in hand, Kalmbach quickly set out to deliver the quid pro quo agreed upon by Semer and him, i.e., a "clear understanding in exchange for this contribution that I would arrange for Mr. Semer to be able to see certain individuals within the administration before whom he would be able to plead his case on behalf of his clients."[11]

Haldeman indicated to Kalmbach that such meetings between the White House staff and dairy lobby representatives were acceptable. A week after receiving the cash from Semer, Kalmbach met with John Ehrlichman at the Western White House in San Clemente and received his permission as well. Kalmbach then called Semer and told him to contact Harry Dent of the White House staff for an appointment. Semer did so and arranged for himself, Parr, and Nelson to meet at the White House on August 19 with Dent and Gleason.[12] At this meeting they generally described to Dent the dairy lobby's substantive problems and again mentioned how they wanted to support the president. No one was crude enough to bring up the recent $100,000 contribution. However, as Nelson saw it, even $100,000 did not take the dairy lobby out of the woods, as they were still thought of as "Humphrey people and not pro-Nixon."[13]

Immediately following that first White House meeting, Harold Nelson sent a letter to Harry Dent inviting the president to attend the annual meeting of Associated Dairymen, Inc. (a federation composed of AMPI and Mid-Am) where an anticipated six thousand farmers would be in attendance. The matter was mentioned in several meetings with White House officials, including one with White House Special Counsel Charles Colson but, in spite of their efforts, AMPI was able to secure only the appearance of Clifford Hardin, then secretary of agriculture. The importance of having the president at the convention was symbolic. Nixon's appearance would have strengthened the dairy lobby's credibility and demonstrated it had gotten its foot in the door of the administration. Nixon, however, did call Nelson personally before the convention, wished him the best of luck, and asked Nelson to express his regrets to the assembled multitude.

In the fall of 1970, Nixon even arranged to meet personally with Nelson and Parr and again expressed his regret at missing their annual meeting. He told them he had received "good reports" on their activities, and they pledged their support to him for his re-election.

During this time, Nelson and other dairy lobby leaders were beginning to discover some of the problems that went along with buying access to Richard Nixon's White House.

In late 1969, one-time White House aide to President Johnson, Washington attorney W. DeVier Pierson, an AMPI lawyer since early 1969, told Harold Nelson and other co-op leaders that the $100,000 contribution—being a contribution exceeding $5,000—must be reported to the clerk of the House under the then applicable Federal Corrupt Practices Act. But there was a catch to this. Since the contribution was in cash and exceeded the legal $5,000 limit, if it was reported, TAPE would be admitting a violation of the Federal Corrupt Practices Act. Faced with this dilemma, Nelson decided to hide the fact that it had come from TAPE funds.

The first part of the "cover-up" plan was to reimburse TAPE for the Kalmbach contribution and thereby alleviate the need to explain any discrepancies in existing TAPE funds. This reimbursement was to be made through Bob Lilly, who, since he made less than $40,000 a year, would get the necessary funds via a personal loan from Citizens National Bank of Austin, Texas, whose chairman of the board was still AMPI lawyer, Jake Jacobsen.

The second part of the cover-up scheme was to recoup the $100,000 in $10,000 lumps from lawyers DeVier Pearson, Joe Long, Frank Masters, Stuart Russell, James Jones, Richard McGuire, Clifford Carter, and P.R. consultant Ted VanDyk. Moreover, $5,000 each was to be obtained from AMPI employees Lilly, Parr, J. G. Anderson, and Leo Suttle, twelve benefactors totaling $100,000. The recoupment was to be laundered through billings to AMPI from the eight AMPI attorneys and VanDyk of $20,000 each, a sum which would cover their $10,000 contributions plus excess income taxes. The $5,000 each from the four AMPI employees would be made up to them with $5,000 expense account advances. The total cost to AMPI of covering up the TAPE $100,000 contribution to Nixon would be approximately $180,000.[14] The original plan was subsequently modified so that the four AMPI employees would not participate, but the outside lawyers and VanDyk would. Either way, the plan involved illegal diversion of AMPI's corporate assets for political purposes.[15]

To repay TAPE by the end of the year—thus avoiding the reporting requirement—Bob Lilly borrowed the entire $100,000 and protested so vigorously to Nelson about being used in this manner that he later claimed to have almost lost his job. On December 17, 1969, a loan application was made by Lilly to the Citizens National Bank and, in due course, he was lent $100,000. Jake Jacobsen, along with two other members of the bank's Loan Discount Committee, approved the loan, which was secured by a $100,000 certificate of deposit owned by AMPI and pledged on AMPI's behalf by Lilly, even though there was no evidence that Lilly had authority to use AMPI assets in this fashion.

Lilly took the loan proceeds and deposited them into the TAPE account. Because Lilly's pledge of the $100,000 AMPI certificate of deposit may have been defective, a TAPE check dated December 19, 1969 was written to the bank to purchase for TAPE a $100,000 certificate of deposit. This certificate was then backdated and pledged as security for Lilly's loan to substitute for the questionable AMPI certificate of deposit. Since TAPE's assets were now at full reportable strength, it concealed the $100,000 to Kalmbach by simply not reporting the pledge of the certificate of deposit as security for Lilly's loan. Lilly's sixty-day note, however, was not paid until nearly one year later, the mechanics of the recoupment scheme taking that long to work properly.[16]

AMPI's board of directors has attempted to disclaim responsibility for the illegal use of AMPI corporate funds in the recoupment scheme for the Kalmbach contribution, but their argument is tenuous at best. It is true, from all available accounts, that the AMPI board did not know specifically of the $100,000 to Kalmbach. It is not true, however, that they were unaware of the political use of AMPI corporate funds by Nelson and others. The recoupment scheme of excess billings by AMPI attorneys to cover political contributions was standard procedure with AMPI, and its board knew it. Nelson discussed with the AMPI board "on more than one occasion" that attorneys fees were high because of the laundering of political contributions and reimbursements for taxes. These matters were discussed with the AMPI board at meetings in Las Vegas and Madison, Wisconsin, and the board quite clearly understood that remarks about funnelling attorneys fees through excess billings were not to be included in the minutes. In fact, according to Nelson, "some of the board members actually called on him to do some of those things," while other board members (including John Butterbrodt, AMPI president and chairman of the board)[17] turned in expense accounts to cover their *own* political contributions.

The Nixon White House continued its policy in late 1969 and 1970 of allowing the dairy lobby frequent access to White House staff, initially Jack Gleason and Harry Dent. At all of these meetings, the dairy lobby—usually represented by Nelson, Parr, and one of their lawyers—would discuss the dairy lobby's position on milk price supports, dairy import quotas, and tariffs, and also the dairy lobby's anticipated contributions to Nixon's 1972 campaign.

By the middle of 1970, Gleason had relinquished his role as liaison with the dairy lobby to White House Special Counsel Charles Colson. At their first meeting with Colson, the dairy lobby leaders asked him for a list of committees to receive campaign contributions. At another meeting with Colson, Dave Parr mentioned that there was a $2 million potential for political contributions from the dairy lobby.[18] Since the presidential race was two years away, Parr wanted Colson to know that the dairy lobby had a potential for contributing to congressional campaigns as well as presidential campaigns and that all three political action committees, TAPE, ADEPT, and SPACE, were coordinating their contribution activities.[19]

One of the substantive policy problems inherited by Colson with the dairy lobby involved dairy imports. In his June 1970 memo to Colson giving him responsibility for the dairy lobby, Gleason commented on:

> the possibility of having the president request that the tariff commission take emergency action on dairy imports in a similar fashion to that which Johnson did in July 1967, following the March 30, 1967, request for a tariff commission investigation. I am attaching to this memo a copy of a letter Parr prepared which spells out in some detail what they are looking for. The problem is evidently that since we recommended the tariff com-

mission begin an investigation of dairy imports again, the European importers have begun to dump increased quantities of their product on the market. Parr is cognizant of the line of the president's last address on the state of the economy regarding the possible need for increasing all imports to offset inflationary pressures, but that, of course, in no way lessens his interest in achieving the above. . . . Over to you.[20]

Colson had not done much on the problem by October 1970, when the tariff commission issued its long-awaited report. On the same day it was issued, AMPI attorney Marion Harrison wrote to Colson describing the difficulties faced by the dairy lobby because of unrestricted dairy imports and urging action in time to make a favorable political impact on the 1970 congressional races:

> The longer the delay, the more people in the dairy related industries will learn that the President is not following the well-documented recommendations of the Tariff Commission.[21]

Colson—and Nixon—ignored Harrison's plea, and the November elections passed with the recommendations of the tariff commission still not acted upon. By this time, the dairy lobby was sufficiently concerned about Nixon doing nothing on import quotas that Marion Harrison's law partner, Pat Hillings, a former congressman from Nixon's old district, wrote personally to the president on December 16, 1970, and openly reminded him that the dairy lobby had contributed about $135,000 to Republican congressional candidates in 1970 and was working with Herb Kalmbach and Tom Evans, one of Nixon's old New York law partners, in setting up "appropriate channels" for the dairy lobby "to contribute $2 million for your re-election." Apparently believing that this crude reminder was sufficient to attract the president's attention, Hillings then proceeded directly to the dairy import question and complained that "it never took the Democrats this long" to act favorably in the matter of quotas on behalf of the dairy industry, and that LBJ had "even imposed quotas before he received the Tariff Commission's recommendations."[22]

While Nixon never saw it, Colson was not pleased by Hillings' heavy-handed letter or by Hillings' partner, Marion Harrison. For one thing, Colson did not like the dairy lobby contributing to Democrats. Prior to the election, he had written a scathing memo to Nixon's old political mentor, Murray Chotiner, who was then in the process of making arrangements to leave the White House and join Harrison's law firm:

> Would you please check with your friend, (Marion) Harrison, and tell him if he wants to play both sides, that's one game; if he wants to play our side, it's entirely different. This will be a good way for you to condition him before we put the screws to him on imports, which we are about to do.[23]

Upon the receipt of Hillings' letter to Nixon, Colson promptly sent another memo to Chotiner:

> Your friends, Harrison and Hillings, have just about run out of string with Henry Cashen and me. They are personally abusive —particularly Harrison—not only to the two of us but to the secretaries in this office and are making impossible demands. They continually go around us. They have told us that we cannot under any circumstances talk to their principals. Their clients, of course, continue to call us and in an effort to be helpful to Harrison and Hillings we refuse to take the calls.
>
> They have so muddled up the present dairy import situation that I almost think there is no way that we can help them. It is, believe me, an incredible situation. I practiced law for ten years in this city and wouldn't think of treating a messenger from GSA the way these guys think they can order the White House around. Frankly, in view of the relationship with the dairy industry that is involved, I think these guys are simply too dangerous to deal with and that they should either be put in their place or cut out of the act altogether. . . . In sum, they are very, very bad news.
>
> Unless this situation can be straightened out, I intend to talk to the dairymen directly . . . and simply let them know that we will deal with them on any problems but we cannot deal with their lawyers—or at least in the way which their lawyers have been acting.[24]

Colson's animosity toward Harrison and Hillings apparently prevailed within the White House because when Nixon finally ended the suspense and issued a proclamation setting import quotas, they were at a level lower than recommended by the tariff commission. Although privately disappointed at the lower level, John Butterbrodt, chairman of AMPI, announced in a press release that

"President Nixon's decision is a step toward more stability in our market that will be remembered and appreciated by farmers."[25]

While Colson may have been displeased with Harrison's personal treatment of him and his people, it was this same Marion Harrison whom Colson subsequently used to acquire from AMPI the untraceable cash necessary to fund the White House break-in at the Los Angeles office of Daniel Ellsberg's psychiatrist. Ellsberg was the former Defense Department official who had incurred the wrath of the Nixon Administration by obtaining and arranging for the publication of the "Pentagon Papers" which clarified previous presidents' roles in the Vietnam war.

Colson asked Harrison in 1971 to have the dairy lobby make a "special" $5,000 contribution to "People United for Good Government," one of the many fund-raising committees established by the National Republican party to receive political contributions. Its treasurer was George D. Webster, a man described by Harrison as "a very prominent Washington lawyer."[26] Harrison agreed to do so and as a result of his recommendation, TAPE promptly issued a $5,000 check. Harrison took the check personally to Webster who, acting on Colson's orders, eventually gave $5,000 in cash to Joseph Baroody.

Joseph Baroody was employed in the Washington public affairs firm of Wagner and Baroody and represented clients "affected by federal government actions." In a cryptically worded memorandum in 1970, Colson had informed Haldeman that "I have arranged for a public relations firm here in town [Wagner and Baroody] to do things on our behalf. . . . Some of our friends [the dairy lobby] have retained this outfit and essentially given them the financial resources to do things for us. The funds can't be used for direct media expenses because these would not be deductible for the public relations firm. They can be used, however, to put people on their payroll, pick up travel expenses . . . and provide a perfect cover for us."[27]

Joseph Baroody had been acquainted with Colson for several years and, in August or September of 1971, Colson told him of the White House's urgent need for $5,000. Baroody gathered this amount from personal and office funds and the day after the call took $5,000 in cash in an unmarked envelope to Colson's office in the Executive Office Building (EOB) across from the White House. Colson then mysteriously told Baroody to take it to another office in the EOB and give it to someone who was expecting it. The office to which Baroody delivered the money belonged to Egil Krogh, John Ehrlichman's chief accomplice in plotting and carrying out the Ellsberg break-in.

Several weeks later, Colson phoned Baroody and told him to go to George Webster's office where he would be repaid for the $5,000 he had advanced to Egil Krogh. Webster was not in when Baroody arrived at his office, but Baroody identified himself to the receptionist, who handed him an unmarked envelope containing $5,000 in cash, a transaction made possible by TAPE's earlier $5,000 contribution to Webster's committee.

The circle was complete. The White House "Plumbers" had their money; the Ellsberg break-in proceeded; the money to fund it was appropriately laundered via an otherwise legal contribution from the dairy lobby, whose own PR man Baroody had advanced the funds in the first place.

As for the dairy lobby, it had long known the high price of buying "access" to the Nixon White House. It had not known of the bargain it was receiving—for the same money, it had also bought vicarious access to the psychiatrist's office of one of the men at the top of Richard Nixon's enemies list. Far worse was in store for the dairy lobby, however, than the $5,000 it had so innocently and willingly contributed at Charles Colson's behest in 1971.

Part 3

The Abuses of Power II: Milk and Watergate

Uh, I know, . . . that, uh, you are a group that are politically very conscious. . . . And you're willing to do something about it. And, I must say a lot of businessmen and others . . . don't do anything about it. And you do, and I appreciate that. And, I don't have to spell it out.

> President Richard M. Nixon, transcript
> of meeting with milk-producer cooperatives,
> March 23, 1971.

They are tough political operators. This is a cold political deal.

> President Richard M. Nixon, transcript
> of meeting with John Connally,
> March 23, 1971.

6

Nixon, Nader, Price Supports, and the Road to Impeachment

Despite their private disappointment over Nixon's failure in December 1970 to raise dairy import quotas as high as the tariff commission had recommended, leaders of the dairy lobby were optimistic as they faced the new year. In March, 1971, the secretary of agriculture, Clifford Hardin, was scheduled to announce the milk price support level for the coming year, and Richard Nixon had at last agreed to a large formal meeting with all the leaders of the dairy lobby late in March. Since the dairy lobby had been actively pushing for an increase in the price support to 85 percent of parity, they had no reason to suspect the administration intended to disappoint them.

The milk price support is tied to a "parity" level based upon the golden age of farming—the years 1910 to 1914. Simply stated, this means that a 90 percent parity price support should assure a dairy farmer buying power equal to 90 percent of that which he had in the golden age. The secretary of agriculture makes this price support decision, although he was restricted in 1971 to support levels between 75 percent and 90 percent. The government insures that prices do not fall below the support level by utilizing the Commodity Credit Corporation (CCC) to purchase milk products (butter, cheese, and dry milk) in the open market when the prices fall to the support level price. By offering to buy all excess production of these products at the support price, the government is able to maintain milk prices at or above the support price level.[1]

During the period leading up to the secretary's decision scheduled for March, Dave Parr and Harold Nelson made a point of regularly seeing Secretary Hardin and Under Secretary Phil Campbell to express their views on the 85 percent level. Economic studies prepared by staff economists working for dairy co-ops were also furnished to USDA officials to convince them of the technical merits of an 85 percent parity level.

The dairy lobby was busy on Capitol Hill as well. It persuaded Hubert Humphrey to give a speech on the Senate floor criticizing the Nixon Administration's anticipated price support level of 85 percent as inadequate and urging that it be set at 90 percent. Senator Walter Mondale of Minnesota was persuaded by the dairy lobby to give a similar speech the next day.

Dairy lobby attorney Marion Harrison secured a copy of Humphrey's speech and sent it on March 11 to Charles Colson along with an accompanying note which said: "You think 'political' as I do so you can understand the enclosure (a speech of Senator Humphrey on behalf of the dairy farmers)." Harrison went on to refer to Nixon's recent import quota decision as being "more or less" what the dairy lobby had wanted out of the Nixon Administration but added that:

> ... the strain on all of us and the delay are so great that the watermelon does not seem as juicy when the dairy industry finally bites into it. The last major item the industry will request for some time to come ... is 85 percent parity for the year beginning April 1, 1971. The sooner that gets announced, the sooner somebody like Mr. Humphrey can be silenced.[2]

Even though the dairy lobby was trying as best it knew to buy influence, the Nixon White House made no real effort to affect Secretary Hardin's decision on price supports. By encouraging the $100,000 cash payment of Kalmbach, as well as substantial contributions to Republican congressional candidates in 1970, Nixon was essentially allowing the dairy lobby to place its bets and play the game. He simply did not intend to let them win anything. Nixon himself was not sympathetic to a price support increase. A memorandum from Budget Director George Schultz reporting on a March 5 morning meeting of the president, Secretary Connally, Dr. Burns, and Dr. McCracken shows how closely Nixon was monitoring the price support situation. It notes that "strong pressure to raise dairy prices" originated from the dairy lobby, and that Arthur Burns, chairman of the Federal Reserve Board, "argued strongly for doing everything possible to keep the prices from rising insofar as the consumer is concerned."[3] Schultz also opposed an increase in price supports and the day before had received a memo from his assistant director Don Rice which said that "If we accede to the dairy interest on these . . . issues, the result would be higher budget costs, increased production and surpluses of milk products. Next year at this time, the president would surely be faced with a choice between even larger surpluses or a decrease in the support price, an unhappy choice in an election year."[4] Rice therefore recommended holding the line on the price support level and Schultz agreed.

The dairy lobby's imagined "friend at court," Charles Colson, similarly made no real efforts to persuade Nixon to raise price supports. Colson did send on March 10, only two days before Hardin was to announce the decision, an "eyes only" confidential memo to Nixon's top domestic assistant, John Ehrlichman, recommending that "affirmative action on cheese imports" be taken at the same time the parity levels were announced, because of "the obvious political support we discussed" and the fact that the action would enhance the dairy lobby's ability to "sell" the dairy farmers on the unchanged parity levels.[5]

The leaders of the dairy lobby were not aware at the time that parity levels were going to be unchanged and that they would have anything to "sell." They thought the fix was in. Imagine, therefore, the collective shock they suffered when Secretary Hardin announced on March 12, 1971 that the milk price support level for the coming year would stay right where it was—*below* 80 percent of parity. Hardin said that dairy farmers "knew from past experience that they do not benefit when dairy production substantially exceeds demand and excessive surpluses pile up in government warehouses. We must avoid this."

The dairy lobby felt betrayed. It was as if all its work on the

Nixon Administration during the past 2½ years had been for naught and Nixon had forgotten the $100,000 in cash to Herb Kalmbach, the $135,000 to GOP congressional candidates in 1970, and the $2 million pledge for his re-election campaign.

But Nixon had not forgotten. He had simply taken a calculated political risk. The dairy lobby had not complained the previous December when he gave it half a loaf on import quotas. Inflation was a big political problem for Nixon and growing bigger. An increase in the milk support price would increase inflationary pressures. So Nixon decided to do "the right thing" and "tough out" the political consequences. In attempting to do so, however, he and his political aides simply underestimated both the extent of the dairy lobby's political power in Congress and elsewhere within his own administration.

The dairy lobby promptly launched a massive three-pronged counterattack aimed at Congress, Nixon's newly appointed treasury secretary, John Connally, and the Nixon White House itself. Their goal in Congress was to create a groundswell for at least 85 percent, if not 90 percent of parity. Virtually overnight, the dairy lobby flooded Capitol Hill with over fifty thousand letters urging an increase in parity. Operating from a three room suite in Washington's luxurious Madison Hotel, the dairy lobby succeeded in having thirty bills introduced in the House sponsored by 125 congressmen mandating a full 90 percent parity level. In the Senate, twenty-eight senators introduced legislation calling for an 85 percent parity level.

AMPI legislative director Bob Lilly believed that these lobbying efforts had a significant impact, and that the dairy lobby would have had at least two hundred and twenty-five votes on the bill in the House and well over fifty votes in the Senate. By this time, the legislative route was perceived as so successful by some leaders of the dairy lobby that they did not wish it abandoned, feeling that they could override any veto the president would impose and that giving up legislation was somewhat of a betrayal of those legislators who had risked supporting dairy interests.

Antiadministration feeling was now running high enough among dairy lobby officials that an earlier promise to Colson to attend in strength a March 24 $1,000-a-plate Republican fund-raiser dinner started to weaken. That commitment to Colson for $100,000 worth of tickets was ten times higher than the number of tickets the dairy lobby would normally have bought. Several dairy leaders even talked about a *complete* dairy cooperative boycott of the dinner.[6]

During this same period, the dairy lobby had Jake Jacobsen renew his contact with John Connally. Jacobsen, a long-time political friend of Connally's, had met earlier with him on March 4 to discuss milk price supports and Connally promised to be helpful if he could. He met again with Connally on March 18, explained the extent of congressional support the dairy lobby had for 85 percent or even 90 percent of parity, and explicitly warned Connally that the dairy lobby's financial support to Nixon's re-election would not be forthcoming unless the March 12 price support decision were reversed.

The day after Jacobsen's meeting with Connally, Marion Harrison sent a memo to John Ehrlichman's assistant, John Whitaker, in which he flatly warned that "for political, if no other, reasons, parity must be again set at 85 percent, even if the president has to do it. *The president's name, not the secretary's, is on the ballot.*"[7]

The Nixon White House, however, did not need to be reminded in such an unsophisticated way of the dairy lobby's power. On the same day that Harrison sent him a memo, Whittaker had sent his own memo to his boss Ehrlichman suggesting just how closely the White House was following the dairy lobby's progress on Capitol Hill:

> I think we should have a prompt meeting with Secretary Hardin today. The prime issue is milk price supports. Contrary to what I reported in the 7:30 meeting this morning on a House count they did last night, Hardin is convinced there is a 90 percent chance that an 85 percent parity price support for the milk bill, sponsored by Carl Albert, will pass Congress. The issue is, if it passes, does the president veto it. Currently we are playing a bluff game with the dairy people saying the president will have to veto a milk price increase and [take] credit on the consumer side, but Hardin doesn't think it will stop the bill from passing. He is now of the

opinion that when the dairy meeting takes place with the president next Tuesday, the president should allow himself to be won over and go along with the argument of raising the price of milk to 85 percent of parity.[8]

Whittaker's warning prompted a meeting that same day in Ehrlichman's office, where they discussed congressional price-support pressure, and campaign contributions, as well as the scheduled March 23 meeting between the president and dairy leaders. While the adverse effect on dairy lobby campaign contributions of another negative price support decision was discussed, Ehrlichman's decision at the time was to "tough it out."[9]

Ehrlichman's attitude was possibly caused by a phone conversation he had the day before with AMPI's Dave Parr when he told Parr that "the president wants this congressional effort called off." When Parr responded that "I don't believe you understood me, I can't call it off," Ehrlichman answered, with thinly veiled threats: "You've heard of the Federal Trade Commission, haven't you? . . . . You've heard of the Justice Department, haven't you?"[10]

The threats had no apparent effect, however, for as the March 23 meeting between Nixon and leaders of the dairy lobby approached, the dairy lobby had not publicly budged from its position. Inside the White House there had been no apparent change either.

As with all major meetings in the Nixon White House, the president was prepared for the March 23 session with a comprehensive background memorandum which explained that "the dairy lobby has become very strong [and] lately, has decided, like organized labor, to spend a lot of political money. Pat Hillings and Murray Chotiner, for example, are involved." The memo briefed Nixon about the milk-price support situation and informed him that the White House staff recommendation was that the president appear to hold the line in front of the dairy leaders and await congressional developments over the course of "the next two weeks to see if the Democrats can move the bill."[11] Nixon, however, received contrary advice from John Connally on the morning of March 23, advice which Connally would repeat at length that afternoon for the benefit of Nixon's advisers and the White House taping system.

Leaders of the dairy lobby present with Nixon on March 23 included Harold Nelson, John Butterbrodt and Dave Parr from AMPI, Paul Alagia from DI, Gary Hanman from Mid-Am, and several others. Clifford Hardin and Phil Campbell were present from USDA.[12]

After shaking hands and introducing himself, Nixon told the group he was sorry to keep them waiting but that he had been talking to Connally on the phone and that Hardin and he had talked about an agricultural matter. The president then recognized the dairy lobby's political consciousness and said

> These days it is sort of unfashionable to talk patriotism and the rest. . . . I still do. Now this group is . . . being sort of a bedrock, the heartland, as we call this America . . . Agriculture has done an enormously effective job. . . . You stood for those things that are deeply needed in this country today. . . . Perhaps you are a relatively small group—I just want you to know that in this office . . . that kind of commitment, that kind of support above partisanship, this is something I am deeply grateful for.

Nixon then went on, in a direct reference to the dairy lobby's campaign contributions and resulting political influence, to say:

> Uh, I know . . . that, uh, you are a group that are politically very conscious. . . . And you're willing to do something about it. And, I must say a lot of businessmen and others . . . don't do anything about it. And you do, and I appreciate that. And, I don't have to spell it out."[13]

After the warm introduction, Nixon gave the meeting over to Hardin, and the subject of price supports was discussed in substance. At one point, Nixon said that "we have to look at what's best for the man that's pulling the teats out in the farm, if I can use that old expression." To which some unidentified wag replied, "You do it with machines." Amidst the laughter, Nixon added, "I know something about that," telling them that his mother and father were from Indiana and Ohio and that upon their retirement, "they went back to the farm and so I have a little bit of agricultural background" (obviously antedating machine milking).[14]

The president indicated an interest in what an increased price support level would do to the economy within a year and focused on the ability of the milk groups to control their own production, which the cooperatives said they could do. President Nixon informed them that he did drink a lot of milk and questioned the cholesterol factor. Cholesterol, said Nixon, "is related to stress. . . . You'll have a test made one week . . . above normal, and next week . . . it will be below." And to further buttress his ties with the milk producers, the president criticized the concept of sleeping pills and said that when faced with insomnia "the best thing you can do is milk . . . just a glass of milk. . . . It could be warm. It could be . . . tepid . . . or it could be cold, but it has a certain soothing effect." And advised the dairy people to "get started on that," offering it with a laugh as "my marketing picture."[15]

In a conversation between the president and Dave Parr, the president had asked how the super cooperatives were brought together. After hearing Parr's response, Nixon commented "that's quite an achievement." In the midst of the candid discussion of the dairy lobby's economic power which ensued, someone commented, "Don't say that while I'm sitting here," to which Nixon responded "Oh, I won't go that far. [laughter] Matter of fact, the room is not taped. [laughter] Forgot to do that, [laughter.]."[16]

Nixon clearly made a favorable impression on the dairy leaders during the meeting. Paul Alagia of DI thought that the meeting with the president was "nice" and appreciated the time Nixon took to respond to all the farmers' questions. When he left the meeting, John Butterbrodt of AMPI believed Nixon was going to change his mind. Only Dave Parr was not optimistic and still believed that legislation was their best hope.[17]

Parr turned out to be wrong and, in retrospect, Nixon had probably in his own mind rejected his staff's advice by the time he met with the dairy leaders. In any event, Nixon's decision was confirmed at a meeting he had in the oval office later that afternoon with Connally, Hardin, Schultz, Ehrlichman, Whittaker, Campbell, and Rice. Connally was the dominant figure in the meeting, and Nixon co-operated with him, almost as if they had choreographed the scene in their phone conversation earlier that day:

Nixon: How about the politics? Can you—

Connally: Uh, I'm not trying to talk about it or discuss at any great length the, the economics of it, but as far as the politics are concerned—looking to the 1972, it, uh, it appears very clear to me you're going to have to move, uh, strong in the midwest. You're going to have to be strong in rural America, uh, and particularly that part of the country. Now, there are a lot of things that you can't do, uh, with respect to farmers. They're almost, uh, beyond help at this point. Uh, they feel like they are. They don't feel like anybody's trying to help them. Uh, everytime they turn around, they hear somebody talking about, wanting to increase imports on beef from Australia to—in behalf of the consumer . . . These dairymen are organized; they're adamant; they're militant. This particular group, AMPI, which is the American Milk Producers Institute or something, uh, represents about forty thousand people. Frankly, they tap these fellows—I believe it's one-third of one percent of their total sales or $99 a year whichever is. . . .

Nixon: Like a Union.

Connally: Oh, it's a checkoff. No question about it. And they're meeting, and they're having meetings. They have them a Sabreliner airplane, and they just travel from one part of the country to another . . . they're amassing an enormous amount of money that they're going to put into political activities, very frankly. . . . They've got . . . a legitimate cause. I wouldn't recommend that you . . . do that if it didn't have any merit to it. . . . They're doing some things that I think are a little strong armed tactics, perhaps, in . . . the organizing . . . but . . . I don't criticize that unless we are prepared to take on business and labor and all at the same time. There's no point denying the farmer what's a practice for the labor; and . . . so . . . I wouldn't judge it on a moral basis. . . .

I'm addressing myself to the narrow aspects, to the political aspects of it. I don't think there's a better organization in the United States. . . . If you can get more help for them [they] . . . will be . . . more loyal to you and . . . I think they've got a worthy case to begin with . . . That being true, I just think you ought to stress the point. And I wouldn't wait until next year. . . . They're going to spend a lot of money this year [1971] in various congressional and senatorial races all over the United States. . . . If you do something for them this year, they think you've done it because they've got a good case and because they're your friend. If you wait until next year, I don't care what you do for them. They're going to say, "Well, we put enough pressure on them this election year, they had to do it." And you, you get no credit for it. So it's *still* going to cost you an enormous amount of money next year, and you get no political advantage out of it.[18]

Others present pointed out to Nixon that Wilbur Mills and House Speaker Carl Albert strongly backed the bill. Connally then listed a few of the states where a veto on this issue could cost the president the 1972 election. When the advice had all been given, Nixon looked up and told them what they had probably known ever since he had asked John Connally to speak:

> Well, it's one of those things where with all you experts sitting around you have to make a political judgment. My political judgment is that Congress is going to pass it. I could not veto it. Not because they're militants, but because they're farmers. And it would be just turning down the whole middle America, where, uh, we, uh, where we, uh, need support. And under the circumstances, I think the best thing to do is just, uh, just, uh, *relax and enjoy it* [emphasis added].[19]

Nixon then designated Phil Campbell as the one to talk to the dairy lobby, tell them of the president's political decision, and reconfirm their attendance at the GOP fund-raising dinner the next day. As the meeting ended, Ehrlichman suggested a glass of milk to the group, "Drink it while it's cheap."[20]

Campbell promptly called Dave Parr and Harold Nelson, informed them of Nixon's tentative decision to raise price supports, and told them not to "boycott" the GOP dinner the next day. And just in case the dairy lobby missed the message, Murray Chotiner (acting on instructions from Ehrlichman) also called Nelson on the evening of March 23 and told him to be available for a meeting the next night with himself and Herb Kalmbach to reaffirm the dairy lobby's $2 million dollar pledge to Nixon's re-election campaign prior to the announcement of the price support reversal then scheduled for March 25.

The dairy lobby did have a problem with all this. Because of its disappointment with the earlier Nixon decision not to move on price supports, it had not fulfilled its pledge to Colson to purchase $100,000 worth of tickets for the March 24 GOP fund raiser. Time was now running out. The dinner was less than twenty-four hours away.

Hanman of Mid-Am had gone to bed, and Alagia of DI had

flown to Chicago after the meeting with Nixon. That did not deter Parr. He roused Hanman from a deep sleep in his Washington hotel: "Progress is being made," said Parr; "we should go ahead and go to the dinner." Hanman agreed.

Paul Alagia of DI was more of a problem. No one knew where to reach him in Chicago so they called his wife at Alagia's home in Louisville and told her that Nelson and Parr wanted to meet Alagia in Louisville as soon as possible.

Parr, Nelson, Hanman, and Bob Lilly took off that night in AMPI's private jet for Louisville, where they were joined the next morning at 4:00 A.M. by Alagia, who had flown in by private jet from Chicago. When questioned subsequently by the Senate Watergate Committee investigators as to whether this was an unusual meeting, Nelson (presumably with a straight face) responded that "I would say a meeting at midnight at an airport is not unusual ... not at all an uncommon thing to do with [Alagia]."[21]

After Alagia arrived, they all sat down on benches in the lobby and talked about the Republican dinner and their attendance. It was decided that it would make a better impression if all three political trusts purchased tickets. However, for Mid-Am's trust, ADEPT, to purchase "a significant amount of tickets" also, a loan to ADEPT from either DI's trust, SPACE, or AMPI's trust, TAPE, would be required.

Nelson pressed Alagia for SPACE to make the loan to ADEPT and believed that the urgency of having ADEPT in on the initial contributions was one of the things which merited the night flight to Louisville. Nelson also thought that SPACE should be responsible, along with ADEPT, for at least one-third of the dairy lobby's $2 million pledge.

Alagia refused to make any specific commitment on behalf of SPACE that night and thought of the meeting as a "last minute effort to put some pressure on a guy." Alagia told the others that he was going to leave any decision to Ben Morgan, who was soon to be his successor at Dairymens, Inc. As Alagia recalls the meeting, he was ready to do "anything within limits to politely excuse myself and get out of there. I didn't appreciate them descending on me."[22]

Alagia may not have appreciated it, but DI came through. The result of the 4:00 A.M. meeting in Louisville was that later in the morning of March 24, 1971, SPACE contributed $25,000 and ADEPT pledged $50,000 to Republican campaign committees.

Alagia, after leaving the airport gathering and going home to sleep for several hours, had arrived at DI offices the next morning and informed Morgan and Moser of the new development. They all agreed to make the contribution. As a result, Jim Mueller, SPACE trustee, had five checks for $5,000 each to five Republican committees taken by plane to Washington and delivered to Marion Harrison.[23]

Having sent a messenger with DI's $25,000, Alagia did not attend the Republican fund-raising dinner at the Washington Hilton on March 24, but both Nelson and Parr did, along with Herb Kalmbach and Murray Chotiner.

Kalmbach had been told by Ehrlichman the day before to meet with Nelson and Chotiner in his room after the GOP dinner. He knew from Ehrlichman that the $2 million pledge was to be "reaffirmed" and that he had to obtain the reaffirmation prior to the announcement of the reversal of the price support decision the next day. Even though Kalmbach noted the meeting on his personal log for 11:00 P.M., he inexplicably went back to his suite at the Madison Hotel after the dinner and fell asleep. Nelson and Chotiner went over to the Madison, and Chotiner tried to reach Kalmbach several times on the house phone. When the three finally got together, Kalmbach was in pajamas.

After some pleasantries and discussion about the difficulties the Nixon White House was having in establishing a sufficient number of campaign committees to receive contributions, the meeting got down to business. Kalmbach specifically understood that the price support decision would be announced the following day and that the dairy lobby people "were in fact reaffirming the pledge" of $2 million because of that. Kalmbach recalls that a goal of $90,000 a month was discussed, and Nelson did not object.[24]

Now it was the turn of the bureaucrats at the USDA to be shocked when they learned the next day that Secretary Hardin intended to reverse the March 12 decision they had all agreed upon and raise milk price supports to 85 percent of parity. Sidney Cohen, chief of the dairy branch of Commodity Operations Division in the Agricultural Stabilization and Conservation Service, was the person responsible for the initial preparation and recommendation of the dockets for the March 12 decision. On this occasion, Cohen was not asked to prepare a draft of the press release or docket for the March 25 reversal announcement, even though it would have been "normal practice" to submit an amended docket to the board of directors of the Commodity Credit Corporation before the announcement. Cohen was simply told to prepare the amended docket after the March 25 announcement and to keep it as simple as possible. He was not consulted in any manner about the reversal and never had any knowledge of the political or economic factors leading to the decision.[25]

One of Cohen's aides expressed "shock and unhappiness" over the decision, and an economist for the Agriculture Department, Anthony Mathis, stated flatly that "those who worked on the original docket were 'horror stricken'... there was no technical reason for the reversal."[26]

The dairy lobby was delighted by Hardin's actions and, having been burned once by the administration, seemed to take its commitment to contribute to Nixon seriously after the March 25 price support reversal. Within a week, Marion Harrison received from the White House and sent to Nelson the names of one hundred committees to which AMPI (through TAPE) contributed $2,500 each, a total of $250,000. By September of 1971, $235,000 had been contributed to a total of ninety-three committees, seventy-five receiving money from TAPE and eighteen receiving money from the other two dairy trusts. By January 1972, the total had increased to over $325,000.

Letters from Mid-Am leaders to their members bluntly emphasized the importance of dairy contributions to the price support reversal. Writing a few days after the March 25 price support decision, Gary Hanman told a Mid-Am member that the reversal was procured by Mid-Am and AMPI "with some help from DI," but that:

I can assure you that the TAPE and ADEPT programs as well as SPACE . . . played a major part in this administrative decision. This just proves that a minority, regardless of its number, if it is well organized, dedicated, and adequately financed, can prevail.[27]

The president of Mid-Am, William Powell, bluntly told a member in a letter that:

The facts of life are that the economic welfare of dairymen
does depend a great deal on political action. . . . I've become increasingly aware that the sincere and soft voice of the dairy
farmer is no match for the jingle of hard currencies into the campaign funds of . . . politicians by . . . labor, oil, steel, airlines and
others. . . .

On March 23, 1971, . . . I was in the cabinet room of the White
House across the table from the president of the United States, and
heard him compliment the dairymen on their marvelous work in
consolidating and unifying our industry and our involvement in
politics. He said, "You people are my friends and I appreciate it."
Two days later an order came from the U.S. Department of Agriculture increasing the price support for milk to 85 percent parity,
which added from $500 to $700 million to dairy farmers' milk
checks. . . . Whether we like it or not this is the way the system
works.[28]

In the ensuing months, the Nixon Administration had reason to
regret the candor of William Powell and Gary Hanman. Others
were listening to the rumors floating around Washington, including
Washington attorney William Dobrovir, a friend of Ralph Nader.
On January 25, 1972, Dobrovir filed a lawsuit on behalf of Nader
in the United States District Court for the District of Columbia,
alleging that the Nixon Administration had "improperly and unlawfully" reversed its 1971 milk price support decision in exchange
for large campaign contributions from the dairy lobby. While it
seemed a harmless annoyance at the time to the Nixon Administration, it was one of the first formal steps which would eventually
lead to the impeachment of a president.

By August 1972, pretrial discovery in the Nader lawsuit had
uncovered the Hanman and Powell letters to Mid-Am members
claiming their campaign contributions were responsible for the
milk price support reversal. In December, 1972, Nader's lawsuit
again hit the headlines with deposition testimony from Washington, D.C., public relations executive Robert Bennet, who admitted
he had established one hundred and fifty secret fund-raising committees in support of Nixon's re-election effort so that large contributors could avoid gift taxes by splitting up their contributions
among many committees. Bennet acknowledged that the dairy
lobby had contributed $2,500 each to the one hundred and fifty

committees in July, August, and September of 1971. Bennet also conceded that he had received help on the legal details establishing the committees from attorney Thomas Evans, the managing partner of Nixon's old New York law firm.[29]

More embarrassment followed for the Nixon Administration in January of 1973, when George Mehren, in a deposition in the Nader suit, revealed that Herb Kalmbach had asked "quite unequivocally" for campaign contributions from the dairy lobby and then attempted to shut off these campaign contributions when AMPI sought to make the contributions public.[30] The Nixon White House referred all press inquiries that followed to Kalmbach, who was unavailable for comment.

Harold Nelson was the next dairy lobby figure to point a finger at the Nixon Administration. In a deposition in the Nader lawsuit on May 22, 1973, he testified that it was the White House and not the Department of Agriculture which made the decision in March of 1971, to raise the milk support price. Nelson candidly conceded that before he and other dairy lobby leaders had met with Nixon on March 23 in the White House, he had been unsuccessful in persuading Agriculture Secretary Hardin on the merits of the dairy lobby position.[31] A few weeks later, John Dean, the former Nixon White House counsel, claimed that Nixon knew that the purpose of the dairy lobby contributions was to get an increase in the milk support price. According to Dean, Nixon was "personally aware" of the 1971 campaign contributions from the dairy lobby.[32]

Meanwhile, Nader's attorney, Dobrovir, was keeping up the pressure in his lawsuit and had demanded production of almost 100 official White House documents relating to the March 25, 1971 price support increase. On July 11, 1973, Leonard Garment, then acting counsel for President Nixon, filed an affidavit in federal court in Washington refusing to produce the documents requested by Dobrovir on the grounds of executive privilege. Nixon claimed that he was immune from producing the documents because the release of the evidence "would be injurious to the public interest." As an afterthought, Nixon's lawyers also asserted that the lawyer-client relationship between Nixon and his legal staff prohibited the disclosure of the documents.[33]

Chief Judge William B. Jones of the U.S. District Court for the District of Columbia was not impressed by Nixon's plea. On July 27, he issued an order directing the White House to submit its files on the 1971 price support increase to him. While not committing himself on whether the documents would be eventually turned over to Nader and his lawyers, Judge Jones specifically rejected Nixon's claim that he had an overriding privilege to shield the internal documents from the court's scrutiny. Jones ruled that if his personal examination of the documents demonstrated that the price support increase was made in exchange for campaign contributions, the documents could not be excluded as evidence in the Nader suit and would be turned over to Dobrovir.[34]

On August 15, Nixon advised Judge Jones through his lawyers that he intended to appeal the court's order to turn over the presidential documents on the milk price support decision, again asserting that inspection of the documents by Judge Jones would create a "substantial breach in the constitutional separation of powers." Nixon's lawyers asked Jones to delay implementation of his order, pending appeal. Nixon even managed to enlist the support of Watergate Special Prosecutor Archibald Cox in his dispute with Judge Jones. Cox had Phillip Lacovara of the Watergate Special Prosecution Force appear before Judge Jones on October 16, 1973, to support Nixon's request that Judge Jones stay enforcement of his order regarding the documents, pending appeal. Lacovara argued that the issue of executive privilege asserted by Nixon in the Nader case could better be resolved in the Watergate tapes case then pending before Judge John Sirica, and that the issues in that case and the Nader case were "sufficiently similar" that they could be settled in one case. While not retreating from his earlier ruling that the documents must be turned over to him for his inspection, Judge Jones nevertheless granted a stay of his order pending a hearing by the Court of Appeals on the issue of executive privilege raised both in the Nader case, as well as in the Watergate tapes case being argued by the Watergate Special Prosecution Force before Judge John Sirica.[35]

When the case was heard by the U.S. Court of Appeals for the District of Columbia, Nixon had his lawyers propose a compromise. Nixon attorney Leonard Garment offered to permit private judicial inspection of some of the confidential Nixon Administration memoranda on the milk price support decision. Garment explained that Nixon had authorized this offer in order to narrow the issues in dispute. Garment continued, however, to insist that the most sensitive documents would not be produced. These sensitive documents fell into four categories: (1) internal staff communications relating to views of members of Congress; (2) memoranda on the mechanics of scheduling meetings with, or appearances by, Nixon before dairy lobby officials; (3) briefing papers prepared for Nixon on his meetings with dairy lobby officials; and (4) memoranda from presidential assistants or presidential counsel regarding milk litigation.[36]

The following week, another attorney for Nixon, Fred Buzhardt, acknowledged that there was a tape recording of Nixon's March 23 meeting with dairy lobby officials and stated that Nixon, once again because of executive privilege, was not about to release the tape. Buzhardt also revealed the existence of a February 1, 1972, White House memorandum on dairy contributions and the Nader lawsuit and also a list of pre-April 7, 1972 contributions to the Committee for Re-Election of the President. Both documents were submitted to Judge Jones for his private review. Nader's attorney, William Dobrovir, was not satisfied and the following day filed a motion with the Court requesting the immediate production of the documents and the tape for his inspection, but Judge Jones deferred ruling on the motion.[37]

Sensing the rising tide of public opinion against him, Nixon next had his lawyers offer a conciliatory gesture to the new Watergate special prosecutor, Leon Jaworski. Nixon, asserting his usual executive privilege claim, had steadfastly refused to turn over to Archibald Cox any of the White House documents on the 1971 milk price support increase or the dairy lobby contributions to Nixon's re-election campaign. On November 29, however, Nixon reversed his previous stand and turned over to Jaworski numerous White House documents on the milk price support decision. Nixon's lawyers noted that Nixon had waived his claim of executive privilege, which he had invoked to deny these materials to Cox, but stated that Nixon was still claiming executive privilege over the milk documents relevant to the Nader lawsuit. His lawyers refused to describe publicly the nature or content of the documents being turned over to Jaworski, and Nixon's assistant press secretary, Kenneth Clawson, stated, in response to reporters' questions, that he did not know why Nixon had chosen to waive executive privilege on these documents for Jaworski but had been unwilling to do so for Archibald Cox.[38]

Despite Nixon's claim on November 29 that he was not waiving executive privilege on the dairy lobby documents relevant to the Nader case, less than a week later, Nixon had his lawyers tell Judge Jones that they would soon deliver to him one of the White House tapes, as well as other documents relating to the milk price support decision. They also told Judge Jones they would turn over another one of the tapes and some additional documents to Dobrovir.[39]

Unfortunately for Nader, after Nixon's lawyer turned over the tape to William Dobrovir, he took the tape home with him and played it for the amusement of a group of his friends at a December 17 cocktail party. Even more unfortunately for Dobrovir, Kevin Delaney, an ABC official, was present at the party. Delaney conveniently produced a cassette player for the occasion and just as conveniently ignored Dobrovir's stipulation that the playing of the tape was "off the record." The subsequent publicity given to the indiscreet playing of the tape afforded the Nixon White House one of its few public relations victories in the long Watergate nightmare. Dobrovir was summoned before Judge Jones and freely acknowledged his "very foolish mistake." Judge Jones agreed to a request from Nixon's attorneys that an order be issued sealing from public view all subpoenaed documents and tapes in the Nader case until they were presented to the court and placed in the public record. Judge Jones refused, however, to order the Nader lawyers to cease talking to the press or making public statements about the case and likewise refused Nixon's request to censure Dobrovir, stating that he was "recognized as an able, honorable and responsible lawyer." Jones suggested that Nixon go to the bar association if he believed any ethical standards had been violated.[40]

This was the highpoint of Nader's suit. It hung on after 1973, but it no longer commanded the same headlines it once had. The drama moved onto a larger stage. Henceforth, headlines on milk came from the Senate Watergate Committee, the Watergate special prosecutor and, in the end, the impeachment inquiry of the House Judiciary Committee.

1974 was not one of Richard Nixon's better years. He started out on January 4 by firing his lawyers, Leonard Garment and Fred Buzhardt, and hiring in their place James St. Clair, a senior partner in the old-line Boston law firm of Hale and Dorr. St. Clair was a well-respected trial lawyer, who had served as an assistant to Joseph Welch, special counsel for the United States Army during the famed Army-McCarthy hearings. St. Clair had been vacationing over the Christmas holiday in Tarpon Springs, Florida, when he received a phone call from White House Chief of Staff Alexander Haig asking him to take charge of Nixon's defense in the Watergate case. St. Clair agreed.

St. Clair's appointment was announced on the same day Nixon told Senator Sam Ervin, chairman of the Senate Watergate Committee, that he would not turn over approximately one hundred documents subpoenaed by the Senate Watergate Committee dealing with political contributions from the dairy lobby. In his letter to Ervin, Nixon noted that "there may be some attempt to distort my position as only an effort to withhold information."[41] Ervin responded that there was "nothing in the Constitution of the United States that gives the president the power to withhold information concerning political activities or information concerning illegal activities."[42] After joining Nixon's staff, St. Clair concurred in Nixon's refusal to turn over the documents.

The following week, on January 8, 1974, Nixon went on the offensive and released his long-awaited White Paper on the milk price support controversy. The White Paper was thorough and covered virtually all aspects of the Nixon Administration's relations with the dairy lobby and the circumstances surrounding the 1971 price support decision. Nixon candidly acknowledged that "traditional political considerations" played a major role in his reversal of the price support decision, but insisted that the dairy lobby's campaign contributions did not have "a substantial influence upon [his] decisions."[43]

Nevertheless, the White Paper contained some embarrassing admissions. At an October 26, 1973, news conference, Nixon had claimed that he always refused to have any discussion about campaign contributions and that he "did not want to have any information from anybody with regard to campaign contributions." Yet the White Paper disclosed that Nixon had been fully aware of the dairy lobby's $2 million pledge to his re-election campaign and that the briefing paper prepared for Nixon prior to his March 23, 1971 meeting with the dairy lobby had reminded him of its plans to make substantial campaign contributions. A second embarrassing aspect of the White Paper involved Clifford Hardin, Nixon's agriculture secretary, who had filed an affidavit in the Nader suit swearing that he had reconsidered his earlier decision not to increase price supports solely "on the basis of statutory criteria," i.e. supply costs and farm income. Nixon's White Paper, however, undercut Hardin and openly conceded that "traditional political considerations" had played a prominent role in Nixon's reversal of Hardin's earlier decision. As a result of this, Hardin became the subject of an investigation by the Watergate Special Prosecution Force regarding possible perjury charges.[44]

Nixon, after the release of the White Paper on milk, resisted furnishing any further documents to the Watergate special prosecutor. Finally, in exasperation, Jaworski wrote to the Senate Judiciary Committee, advising them that Nixon's refusal to turn over a large number of tapes and documents needed in the Watergate investigation, including documents involving the dairy lobby, would retard the scope of his investigation of the political contributions of the milk producers in 1971 and 1972. By that time, however, Nixon was not paying much attention to either the Senate Watergate Committee or the Watergate special prosecutor. The impeachment inquiry by the House Judiciary Committee was getting under way, and House members were as interested in milk as were Leon Jaworski and Sam Ervin.

On April 25, House Judiciary Committee Special Counsel John Doar announced that the committee staff would focus part of its investigation on Nixon's role in the March, 1971 decision to raise milk price supports. The next week, Nixon responded to an earlier April 9 Judiciary Committee subpoena for forty-two taped conversations. In responding, however, Nixon made no reply to the committee's request for additional tapes and documents involving the milk price support decision. When some members of the Judiciary Committee complained about this, Nixon lawyer St. Clair stated that the White House intended to resist turning the additional material over to the committee.[45]

In response, the Judiciary Committee promptly released to the press a copy of the document it had sent to the White House earlier that month, captioned "Justification of Request for Presidential Tapes and Documents Regarding the 1971 Milk Price Support Decision." The "Justification" reviewed the history of the price support decision and then made specific reference by dates and times to conversations it wanted to hear involving four individuals: Charles Colson, whom the committee described as the White House liaison with the dairy industry; John Ehrlichman, Nixon's principal adviser on domestic affairs; John Connally, secretary of the treasury; and lawyer Murray Chotiner. The Judiciary Committee "Justification" went on to state that, in requesting conversations involving these four individuals, it was seeking to determine

> whether any of the conversations in any way bear upon the knowledge or lack of knowledge or action or inaction by the president and/or any of the following officials: Mr. Ehrlichman, Mr. Colson, Mr. Connally, or Mr. Chotiner with respect to a plan or course of action to obtain political contributions from organizations representing portions of the dairy industry in return for influencing official acts by the president or other government officials.[46]

Nixon's refusal to turn over any additional material to the House Judiciary Committee, including the tapes and documents regarding the milk price support decision, was widely perceived as an effort to frustrate the Judiciary Committee's investigation. A press conference by Nixon's lawyer, James St. Clair, on May 8, 1974, did nothing to dispel this impression:

> *Mr. St. Clair:* I think in fairness I should make an announcement to you. I just returned from Mr. Doar's office and I advised him that the president has directed me to inform the committee through him that the president respectfully declines to produce any more Watergate tapes for the committee's use.

And, at the direction of the president, I have also advised Mr. Jaworski that the president has instructed me to press forward on our motion to quash Mr. Jaworski's subpoena.

That is the news for the day. . . .

Question: You said you have been instructed to tell the committee they won't get the Watergate tapes. There are outstanding requests for milk and ITT. Did your instructions cover that as well?

Mr. St. Clair: We have said to them that we think we have provided them everything that is pertinent and that we will continue to examine the matters that they have requested . . . and we will consider producing those from time to time. But as far as Watergate is concerned, the president has concluded, I think rightly so, that that full story is now out and it is time for the committee to start its deliberations.

Question: Mr. St. Clair, you are saying today with respect to the milk issue and the ITT issue that you are still willing to consider that, but at a previous session you had said something to the effect that the name of the game here is Watergate, and the committee might as well just forget about the milk case and the ITT case.

Mr. St. Clair: I don't think those are equivalent statements, but I don't think anyone would seriously contest that the name of the game was not Watergate.

Question: . . .Mr. Wiggins of California, among others, has pointed out that the milk case and to some extent the ITT case involved charges of bribery. The Constitution is very clear on bribery as an impeachable offense. How do you justify, in view of the Constitution, any opposition at all or any reluctance at all to furnish the requested material on the ITT and milk?

Mr. St. Clair: I haven't indicated any reluctance at all. We think we have given them everything. We have given them thousands of documents.

Question: You have a request now from the committee for . . . tapes that include milk and ITT. You said the committee might as well forget about them in a previous session.

Mr. St. Clair: I said in the previous session the principal concern of the American people is Watergate. I don't know that that necessarily means that the committee ought to forget about milk and ITT, and I have indicated here today that we are not in any way indicating an unwillingness to produce more information.

We have produced a great deal of information on both of these subjects. But if there is more that has not been produced that relates to the subject matters they have suggested—and they have been quite specific this time in their request—we are going to get it prepared, present it to the president for his release, and I have every confidence he will release it.

Question: He is going to comply with a request for the 141 [tapes] at it applies to milk and ITT?

Mr. St. Clair: Yes, but believe me, the minor part of the 141 relates to non-Watergate matters. The major part is Watergate. . . .[47]

Despite the carefully couched language denying Nixon's "unwillingness" to produce more information on the milk price support decision, Nixon had decided by then to refuse to produce any more information on milk. Earlier, Nixon's Justice Department had filed a motion in Ralph Nader's lawsuit to keep any further information from being disclosed in that action. On May 14, 1974, Federal Judge William B. Jones granted the Justice Department motion in Nader's lawsuit, stating that "it is highly improper in a civil case for the court to take any action jeopardizing the rights of any defendant," a specific reference to Nixon whom he tactfully refrained from naming.[48]

A week later, Nixon himself admitted how serious his exposure on the milk issue was, when he publicly announced in an open letter to Peter Rodino, chairman of the House Judiciary Committee, that he intended to stonewall the impeachment inquiry on milk or any other subject, and that he was reneging on St. Clair's specific press conference assurance that he would consider supplying additional documents regarding ITT and milk. Nixon flatly told Rodino that no more tapes or documents on any subject would be produced. The seriousness with which Nixon viewed the milk problem is evidenced by the fact he simultaneously had St. Clair send a self-serving letter to the Judiciary Committee's special counsel, John Doar, in response to Doar's letter a month earlier requesting more material on milk. After advising Doar that a review had been made "of the material heretofore furnished you relating thereto," St. Clair added that since "you have already been furnished voluminous documents from the Department of Agriculture and from the White House relating to this matter [and] with tapes of the operative discussions during the course of which the decision to increase the support price was reached," Nixon had personally determined that supplying any more material on milk would not "serve any useful purpose." St. Clair concluded his letter by gratuitously advising Doar that "as you know, the president has published a definitive paper" on the milk price support decision, one which he believes "accurately and completely discloses his participation in the decision. In case you do not have a copy, one is enclosed for your information."[49]

Nixon's adamant refusal to honor the House Judiciary Committee's subpoenas, however, only served to put him in jeopardy on yet another impeachment count. He was advised of this in a May 30, 1974, letter from Rodino, sent to him after the committee voted twenty-eight to ten to warn Nixon of the possibility of the committee's construing his refusal to surrender evidence as impeachable misconduct. The letter also warned Nixon that if he continued to refuse to supply evidence, the committee would "be free to consider whether your refusals warrant the drawing of adverse inferences," i.e., that the tapes and documents being withheld were incriminating.[50]

On June 4, 1974, the day the House Judiciary Committee was scheduled to begin hearings on the price support scandal, the Nixon White House staged a small counterattack by leaking information to the press regarding campaign contributions received by members of the House Judiciary Committee from the dairy lobby. All told, sixteen members of the twenty-eight member committee had received contributions from the three dairy trust funds in amounts ranging from $100 to $11,000. The largest recipient was Representative Edward Mezvinsky, a freshman congressman from Iowa. More surprising, however, was the next largest recipient —none other than Chairman Peter Rodino himself, who received $4,100, even though there was scarcely a milk-producing cow located anywhere in his district.[51]

After that minor embarrassment for the Judiciary Committee, things continued to go downhill for Nixon. On June 25, the committee issued subpoenas for eighteen additional White House tapes bearing on the price support decision and the subsequent dairy campaign contributions. Nixon ignored the subpoenas.[52]

On July 16, Nixon's former attorney, Herbert Kalmbach, testified before the committee and told them what they had long suspected: the dairy lobby had been required to reaffirm its pledge of contributions to Nixon's re-election campaign in the two days between its meeting with the president on March 23, and the announcement of the price support reversal on March 25. St. Clair's unconvincing protestation that Nixon's decision was divorced from the campaign pledges of the dairy lobby fell on deaf ears.[53] Earlier, Charles Colson had testified before the committee and claimed that in 1971 Nixon had promised nothing in return for the dairy cooperatives' pledge of $2 million for his re-election campaign. Kalmbach's subsequent testimony directly refuted this. In a meeting with John Ehrlichman on the night of March 24, Kalmbach said he was told that Nixon had decided the day before to raise price supports but had not announced the decision. Kalmbach testified that he was directed by Ehrlichman to meet that night with Harold Nelson and obtain a reaffirmation of the dairy lobby's $2 million contribution pledge. As the Judiciary Committee's statement of information on the milk price support decision, issued two days later, stated:

> In the deliberations leading to the March 23 decision, there is no evidence that new economic arguments or data with respect to the adequacy of the milk supply were considered. During the president's afternoon meeting on March 23 when the decision was reached, Treasury Secretary Connally, at the president's request, discussed in detail . . . the politics of the decision.
>
> The president was aware of the past financial support from the dairy cooperatives and their pledge of $2 million to his re-election campaign. A memorandum sent to the president on March 22, 1971, reminded him that the dairy lobby had decided to spend a lot of political money. These considerations may also have influenced the decision to increase the price support level.
>
> The committee could conclude from the evidence before it that the president, who is without statutory power to do so, ordered the increase on the basis of his own political welfare rather than the statutory criteria.
>
> Evidence before the committee also suggests that the president

directed or was aware of a plan to secure a reaffirmation of the milk producers' $2 million pledge to his re-election in return for the milk price support decision. The president's refusal to comply with the committee's subpoena has left the evidence incomplete as to whether the milk producer cooperatives' contributions were made with the intent to influence the president's official acts or whether the president acquiesced in their acceptance with this knowledge.[54]

Special counsel John Doar proposed an impeachment count covering Nixon's refusal to turn over subpoenaed tapes and documents covering the milk price support decision, as well as other documents. While language in Doar's proposed count regarding Nixon's refusal was changed in its final form through the drafting efforts of Representative Robert McClory of Illinois, the second-ranking Republican on the Judiciary Committee, the intent was not, and it became the third and final count of impeachment against Richard Nixon voted on by the Judiciary Committee.

The vote on the third count was the closest on any of the three counts which were adopted by the committee, passing by a narrow twenty-one to seventeen margin—two Republicans, Representative McClory of Illinois, and Representative Lawrence Hogan of Maryland, joining nineteen Democrats to support the vote. While it gathered the fewest votes of any impeachment article approved by the committee, it was considered by many observers to be one of the most powerful charges facing Nixon, particularly had there been a trial in the Senate, which might have been more inclined to hold Nixon accountable for refusing to turn over evidence relevant to his guilt or innocence.

The Judiciary Committee had been particularly offended by Nixon's conduct, a feeling which led to its vote on Count 3, because Nixon had by that time turned over to Judge Sirica twenty of the sixty-four White House tapes which the Supreme Court had ruled the week before could not be withheld by Nixon from the Watergate Special Prosecutor. The Judiciary Committee believed they had just as much right to the tapes as the Watergate special prosecutor. After all, argued Representative McClory, if Nixon was to be the "sole arbiter" of what evidence the House Judiciary was to be permitted to have to judge Nixon's conduct, "then how in the world could we conduct a thorough and complete and fair investigation?" To McClory, Nixon's attitude in refusing to turn over evidence was "the prime example of stonewalling."[55]

Within a month of the committee's vote, Nixon resigned.

7

The Attempted Bribery of John Connally

The dairy lobby's role as one of the major players in the Watergate drama did not end with Richard Nixon's resignation. Some of the same events which led to the third impeachment count against Nixon played a prominent role in one of the last acts of Watergate—the 1975 bribery trial of John Connally.

John Connally has been a major figure in American politics for almost twenty years: Lyndon Johnson's protégé, John Kennedy's secretary of the navy, governor of Texas in his own right, wounded in the Kennedy assassination by the famed "magic bullet," Richard Nixon's secretary of the treasury, Nixon's first choice for vice president after Agnew's resignation, the third Nixon cabinet member to be indicted by a federal grand jury on felony charges, and an early contender for the 1980 Republican presidential nomination.

It was 5:22 P.M. on Thursday, April 17, 1975, when John Connally faced the judgment of his peers. The jury filed in, solemn faced, staring straight ahead. Connally looked at the jurors. The jury foreman took a glance at Connally and then faced the bench:

The Deputy Clerk: Will the foreman please rise? Mr. Foreman, has the jury agreed upon a verdict?

The Foreman: We have.

The Deputy Clerk: What is your verdict as to the defendant, John B. Connally, on Count 1 of the Information?

The Foreman: The jury finds the defendant not guilty.

The Deputy Clerk: On Count 2 of the Information?

The Foreman: The jury finds the defendant not guilty.[1]

The court clerk turned and asked the jury if that was the verdict of each and every one of them. "It is," replied the jury in unison. Silence. Connally was motionless. The Judge thanked and dismissed the jury, and the spell was broken. Congratulations started circling around the defense table. Reporters pressed forward and began the questions.

"Governor, are you happy?"

"You bet," said Connally.[2]

Connally hugged his wife, Nellie, and stood holding her hand while responding to further questions:

> I hope as long as I live I never lose the desire to participate in public affairs. I've seen the system work today and it has made me more deeply committed to preserving the system.[3]

Later that evening, a victory celebration took place in the Mayflower Hotel suite of Democratic National Committee Chairman Robert Strauss, attended by lawyers, relatives, and friends. A warm glow permeated the room, the pleasant aftertaste lawyers and their clients savor after a total victory. The mood shifted only temporarily when one of the many calls of congratulations was received. The single man most responsible, after Lyndon Johnson, for raising John Connally to the heights of national and even international prominence, as well as causing him the shame and humiliation of appearing in the dock, was on the phone. He talked briefly to Connally's lawyer, Edward Bennett Williams, and, as usual, when he had nothing of substance to say, talked about the football fortunes of the Washington Redskins. The phone was passed to Connally, and he heard the disembodied voice of Richard Nixon speak of sanity being restored in Washington. When their conversation ended, the celebration continued and the warm glow returned.[4]

Federal Judge George Hart, exercising the power inherent in a lifetime appointment and answerable to virtually no one, had taken the unusual step of refusing to reveal the identities of the jurors.

Within minutes of the verdict, U.S. marshalls kept spectators and reporters away from the jurors, who were led to waiting automobiles and delivered to unknown destinations. Nevertheless, some reporters were able to learn the identities of the jurors and talk with them. Journalist Larry King from Texas talked with some and learned that two of the twelve originally favored conviction, that there were two ballots, and that the holdouts for conviction were eventually convinced to vote for acquittal after reviewing the transcript of the testimony of Connally's chief accuser.[5] Within an hour of receiving that transcript, the jury voted for acquittal. *Newsweek* learned the identity of the jury foreman, Dennis O'Toole, and talked with him: "Our verdict," O'Toole said, "meant not that we had found necessarily that John Connally was innocent, but rather not guilty based on the case presented to us."[6]

The case presented to the jury against John Connally involves a story of two men, personal and political friends, who came to a final confrontation across a federal court room in Washington, D.C.: John Connally and his accuser, Texas attorney Jake Jacobsen, a self-confessed perjurer and political bagman for the dairy lobby.

By the time of his trial, John Connally had come a long way from his college days at the University of Texas, where he worked in the library at seventeen cents an hour and ran for student-body president because it paid $30 a month, and he needed the money. After he graduated from law school, his first political job before World War II had been as secretary to Lyndon Johnson, then a congressman from Texas. Johnson was to serve as Connally's political mentor during the two postwar decades.

During the war, Connally's wife, Nellie, had worked at the Johnsons' radio station in Austin, Texas, while Connally served in the navy. After the war, Connally again worked briefly for Johnson by then a U.S. senator. He returned to Texas in the early 1950s to practice law and make some money. Initially setting up his practice in Fort Worth, he was successful at both— thanks in part to his friend and major client, Texas multi-millionaire Sid Richardson. Says Connally of this period:

I went to Fort Worth and visited Mr. Richardson in his rooms at the Fort Worth Club. We talked most of the night. He invited me to join his organization, and he said: "I can hire good lawyers and good engineers and good geologists, but it is hard to hire good common sense." At the end of our talk he told me, "I'll pay you enough so Nellie and the kids won't go hungry, and I'll put you in the way to make some money."[7]

His work for Richardson kept Connally busy with all the legal problems relating to "the production of oil and gas, and the running of two radio stations, and the running of a television station, and operating cattle ranches of approximately four thousand head on seventy thousand acres, running five drugstores in Fort Worth, various mining interests, housing-development corporations, a carbon-black plant, a gasoline-extraction plant, a suburban acreage development, àn oilfield-servicing company . . . and a hotel."[8]

A prominent role for Connally in politics, however, was not forgotten during this money-making period, only temporarily put aside. In the meantime, he continued to serve as a major strategist for Lyndon Johnson, assisting in Johnson's re-elections to the Senate in 1954 and 1960, the latter campaign coinciding with Johnson's simultaneous election as John Kennedy's vice president.

In 1956, Johnson had introduced a bill to limit price controls on oil and gas, and Connally lobbied for Richardson, although he was not registered as a lobbyist. When a scandal broke involving an abortive attempt to bribe Republican Senator Francis Case of South Dakota, columnist Drew Pearson linked Connally to a lobbyist involved in the scandal. Said Connally at the time to a Senate committee: "I had no part in the incident any more than anybody else who was interested in the oil and gas business."[9]

Up to a point, LBJ was known for rewarding his friends. For John Connally, this meant an appointment as Kennedy's secretary of the navy. When questioned during his Senate confirmation hearings on his role in the 1956 oil and gas lobbying scandal, Connally, who in 1956 had told Lyndon Johnson he was *not* a lobbyist but was merely representing himself, changed his story and admitted he had been working for Richardson: "Again, let me reiterate that I was an employee, I was a salaried employee."[10]

Connally was also questioned about fees from his position as co-executor of Richardson's estate after Richardson's death in 1959. Since Richardson's three co-executors would be splitting over $5 million in fees from an estate which included significant oil and gas holdings, the senators questioned whether the secretary of the navy, given the navy's extensive oil reserves (including Teapot Dome), should receive money so closely tied to the oil industry. This was no problem, said Connally to the Senate Armed Services Committee: "I have had a clear understanding with the executors that . . . I would earn no fees whatsoever during the time I was in government service, and that during my government service I would not collect any moneys due and owing me, even for past services."[11] The next day, Connally was paid $50,000 by the estate.

Connally's career as navy secretary was short-lived but provided Connally with the necessary public prominence to return to Texas in 1962 and successfully run for governor—where his political enemies promptly dubbed him with his mentor's LBJ initials: Lyndon's boy John. Less than a year later, Connally and his wife Nellie were in the tragic motorcade that made it safely to Dealey Plaza and no further. Connally was shot in the shoulder and wrist and maintains to this day that he was not hit by the same bullet which had first passed through President Kennedy. As late as 1973, Connally kept reminders of the tragedy in his home: a large framed photograph of Connally, Kennedy, and Johnson, taken at a political breakfast in Fort Worth the morning of the assassination, and a photograph of the Johnsons visiting the Connallys shortly after the assassination with Connally's arm still in a black sling from his wounds.

Another photograph in Connally's home in River Oaks, Texas, a suburb of Houston, shows Connally being sworn in as Nixon's secretary of the treasury. The photograph carries an inscription from the new boss: "To my good friend John Connally, quarterback for the new prosperity—Proud to have you at my right hand."[12]

After retiring as governor of Texas in 1968, Connally did not return to Fort Worth, but instead went to Houston to practice law. He became a senior partner in the renamed firm of Vinson,

Elkins, Searls, and Connally, much as Richard Nixon lent his name in the 1960s to the prestigious Wall Street law firm of Nixon, Mudge, Rose, Guthry, and Alexander. Within three years, Connally came to Washington as a breath of fresh air, Nixon's glittering prize, easily overshadowing the gray faceless men populating Nixon's cabinet. Nixon was as entranced by Connally as he was with Henry Kissinger. Highly intelligent but graceless himself, Nixon always seemed attracted to those who combined both grace and intelligence. First Kissinger, and then Connally, both of whom had come from humble backgrounds comparable to Nixon's, filled that description.

Jake Jacobsen, a Texas lawyer who had known John Connally for over twenty-five years, had neither Connally's grace nor his intelligence. Jacobsen had been involved in Texas politics since the late 1940s when he had worked for Texas Attorney General Price Daniel. Jacobsen stayed with Daniel after he was elected U.S. senator and subsequently governor. In 1962, he was campaign manager for Daniel in Daniel's unsuccessful attempt to defeat John Connally in the governor's race. While Connally and Jacobsen had known each other politically prior to that time, Jacobsen deftly switched allegiances after the 1962 election and became a political friend of Connally. This friendship paid off in 1965, when he went to Washington as a special legislative assistant to Lyndon Johnson.

Jacobsen served with LBJ until 1967, when he returned to Austin, Texas, and formed a law firm with Joe Long. The new firm of Jacobsen and Long was promptly retained by Associated Milk Producers, Inc. on an annual retainer of $30,000 a year, with an arrangement for separate billings for additional work over the retainer. AMPI was to pay the firm of Jacobsen and Long over $247,000 between January 1, 1969 and April, 1972. In April of 1969, Jacobsen decided to trade on his Washington contacts and also became a partner in the Washington law firm of Semer, White, and Jacobsen. Semer, White, and Jacobsen was similarly retained by AMPI and received over $116,000 from it during the next three years. Jacobsen, therefore, was responsible for bringing in, over roughly a three-year period, legal fees from one client at a rate in excess of $120,000 a year. By Washington or

Texas standards, that is a handsome amount for any one partner to bring in on an annual basis from a single client.[13]

By February 1971, the stage had been set and the two principal characters were ready: John Connally, the prominent and wealthy Texas lawyer, had, in a surprise move, been appointed Nixon's secretary of the treasury. Jake Jacobsen was also a wealthy Texas (and Washington) lawyer pulling down $120,000 a year in fees from the national's largest milk producer cooperative, the most prominent member of the dairy lobby.

John Connally became secretary of the treasury on February 11, 1971. The dairy lobby was pleased with the appointment since one of its main Washington lawyers, Jake Jacobsen, was a close friend of Connally's. At the time, the dairy lobby needed all the help it could get to influence the milk price support decision scheduled for March. Two weeks after Connally's appointment, Jacobsen was on the phone. Could he meet with Connally to discuss a problem involving a client of his? Connally readily agreed and an appointment was scheduled for March 4, at Connally's office.

Jacobsen had called Connally at the behest of AMPI's general manager, Harold Nelson, who asked him to intervene with Connally for assistance in obtaining an increase in the $4.66 price support level then in effect. Prior to seeing Connally, Nelson and Jacobsen met at the Madison Hotel in Washington. Nelson briefed Jacobsen on the dairy industry's $2 million commitment to Nixon's re-election campaign.[14]

Jacobsen met Connally on March 4 for over an hour. According to Jacobsen, they discussed the merits of the price support issue, the vast financial resources and political power of the dairy lobby, the size of the three largest dairy cooperatives, and their checkoff system, which required members to pay $99 per year or one-third of one percent of their yearly gross receipts (whichever was less) to their various milk co-op political action funds. Jacobsen also told Connally of the dairy lobby's $2 million pledge to Nixon's re-election campaign. Jacobsen had learned from Nelson that the Department of Agriculture and the Office of Management and Budget were both against an increase in the price support levels. Jacobsen asked Connally to see what he could do.

Connally agreed and said he would try to be helpful if he could.[15]

Connally, however, offered a somewhat different recollection of the meeting to the Senate Watergate Committee. It was similar to the story Jacobsen had told at the time to the Senate Watergate Committee, except that Connally specifically denied that Jacobsen discussed with him how AMPI's political arm had progressed since they had last talked and claimed that Jacobsen had not mentioned AMPI's political arm at all from early 1969 until early 1971.

After seeing Jacobsen, Connally met with Nixon on March 5 and March 11 and discussed the forthcoming price support decision with him. It did not do much good. On March 12, 1971, Agricultural Secretary Hardin announced the price support level would remain unchanged.

Harold Nelson was not happy. He called Jacobsen to see if another meeting with Connally could be arranged. Jacobsen agreed to do so and to meet with Nelson at the Madison Hotel immediately prior to the session with Connally. Jacobsen claims that he suggested to Nelson that it would be "helpful" if AMPI could pledge additional funds and give credit for that pledge to Connally. Jacobsen did not mention this to Connally, however, when he met with him on March 19 and expressed to Connally the dairy lobby's bitter disappointment at the March 12 decision. According to Connally, Jacobsen told him that the decision:

> was going to create chaos in the milk industry where they were already losing tremendous numbers of cattle and herds and a great many of them going out of business and that they were frankly going to turn to Congress for relief and they had done so, and that they had enormous support and they frankly were going to push for their 85 or 90 percent of parity.[16]

Jacobsen is specifically quoted by Connally as saying:

> I want you to know this is going on because we are not trying to undercut the administration, we are not trying to create problems for you, but we do not think we have been treated fairly and we don't have any recourse except to proceed to try to get congressional relief. We think beyond any question we are going to be successful and we just want you to know this.[17]

Connally admitted he had no argument with what Jacobsen was

doing and understood his position clearly. Connally was certain, though, that Jacobsen never mentioned to him any matters concerning political contributions, the same recollection which Connally had about their March 4 meeting. Jacobsen's recollection was different. He said he told Connally that if the March 12 decision was not reversed, the fiscal support previously pledged to Nixon's re-election would be terminated.

Connally's denial that he talked to Jacobsen about political contributions is suspect because Connally later testified before the Senate Watergate Committee that the *only* person he talked to from AMPI was Jake Jacobsen; that he had not talked to Jacobsen about AMPI's political arm from early 1969 to early 1971; and that he did not talk with Jacobsen about AMPI's political arm or its political contributions at either the March 4 or March 19 meeting. Nevertheless, when the fateful White House meeting of March 23, 1971, rolled around, Connally was remarkably well informed in rather elaborate detail on just how the dairy lobby's political committees operated.[18] Just how well informed Connally was on the political power and organization of the dairy lobby did not escape the notice of the Watergate Special Prosecution Force. In an internal memorandum, the prosecutors commented on Connally's participation in that meeting:

> The next ten or fifteen minutes on March 23, 1971, could fairly be characterized as an intense effort by Connally directed towards eliminating all options for the president other than to raise the price support level. Connally explicitly eschews dealing with the merits of the issue and spends a great deal of time discussing the political power of AMPI and the political consequences of any decision other than an increase. Connally mentions the three co-ops by name; he mentions the number of members each has; he discusses the checkoff system, referring specifically to the $99 a year or one-third of one percent of the gross yearly receipts, whichever is less, arrangement. He tells the president that if the price support level is increased by legislation, that it will be the same drain on the budget as if it were increased administratively, except that Democratic leaders in Congress will get the credit for helping the farmers. Connally says that because of the dairy lobby's considerable financial resources and because of the significant legislators behind that increase, i.e., Carl Albert, Wilbur Mills, he thinks such a bill will pass. He then tells the president

that a veto of the bill will cost him electoral votes in three and probably six states in the farm belt.[19]

Jacobsen called Connally later on March 23, after Connally had talked to Nixon. Jacobsen claimed that Connally gave him advance knowledge of the administration's pending reversal of the March 12 decision, which Jacobsen passed on to Harold Nelson. Connally denied this and claimed he had no advance knowledge of the decision.

On April 28, 1971, Jacobsen met with Connally for half an hour in Connally's office in the Treasury Building at noon.

Connally told Jacobsen, "You know, I was of help on that milk producers thing, and I understand they have some political money. Do you think you can get some of that money for me?" Jacobsen thought he could. Connally denied this conversation occurred.[20]

(A jury found John Connally not guilty of two counts of accepting an illegal gratuity in connection with performing an official act. From all that is publicly known, this verdict was based in large part on the jury's disbelief of Jacobsen's testimony. John Connally, however, was also indicted for a conspiracy to obstruct justice (i.e., a coverup), two counts of perjury, and two counts of giving false testimony. Connally never stood trial on any of these five counts. Connally successfully sought a separate trial on the two bribery counts over the opposition of the prosecutors, who claimed that the conspiracy, perjury and false testimony charges were "inextricably intertwined" with the bribery counts. Yet it was precisely in the area of the conspiracy, perjury, and false testimony counts that the prosecutor's documentary and circumstantial evidence was strongest. The narrative which follows necessarily covers all seven counts of the indictment against Connally, because all seven counts *are* inextricably intertwined. Indeed, Connally's lawyer, Edward Bennett Williams, made no objection to the prosecution introducing evidence on all seven counts at the trial. In the interests of fairness to John Connally as well as of clarity, italics are used in this chapter to denote those portions of Jacobsen's or Connally's respective versions of the facts given in various forums which are either (a) disputed by other sworn testimony, or (b) uncorroborated by other witnesses or documents.

The nonitalic portions of the narrative which follow are, to the best of our knowledge, uncontradicted.)

On the same day, Jacobsen called Bob Lilly in San Antonio and "requested $10,000 for John Connally. [Jacobsen] stated that Mr. Connally had delivered for us on the price support and we were obligated to him for $10,000. [Jacobsen] wanted the $10,000 in cash and he wanted to put it into Mr. Connally's or his box in the Citizens National Bank in Austin."[21]

On May 1, Lilly called Stuart Russell in Oklahoma City and arranged to meet with him in AMPI's offices in San Antonio on the morning of May 3. At the meeting, he explored with Russell the possibility of obtaining the $10,000 in cash from him through AMPI's usual laundering scheme. Russell said that he would do it in this manner but that he had to be paid more than $10,000 in order to compensate him for his increased income taxes. Lilly then discussed with Russell various ways of getting the money without going through the costly laundering scheme, in Lilly's own words, "set up dummy procedures, accounts, a repair account, etc." After concluding his meeting with Russell, Lilly then went to see his boss, Harold Nelson, to discuss the matter with him.[22] Nelson told Lilly that Dave Parr was coming over that afternoon and that Nelson would talk with Parr about it and make a decision as to how Lilly would obtain the money, i.e., either from AMPI's old conduit Stuart Russell, from other attorneys, or by borrowing.[23] From all available accounts, the fundamental question of *whether* to make $10,000 in cash available for the personal use of Nixon's secretary of the treasury was never raised by any of these prominent officials of the dairy lobby.

On the morning of May 4, Nelson got back to Lilly and advised him to borrow the $10,000. That same day, Lilly went over to Austin to the Citizens National Bank and signed Note No. 17266 in the amount of $10,000 and received the proceeds in cash in $100 bills. After leaving the bank, Lilly visited Jacobsen's law office and delivered the $10,000 personally in an envelope. At 4:50 P.M. that afternoon, Jacobsen went over to the Citizens National Bank and signed into safety deposit box 865. *Jacobsen said that he put the money from Lilly in box 865.*[24]

Jacobsen talked to John Connally on the phone on May 7 and

8. Connally's telephone logs confirmed this. *According to Jacobsen, he told Connally that he had what they had spoken about and he was ready to bring it, and Connally said that was fine (Connally denied this).*[25]

On the morning of May 13, 1971, Jacobsen returned to safe deposit box No. 865 at Citizens National Bank in Austin, Texas. *According to Jacobsen, he withdrew the $10,000 which he had placed there on May 4.* Jacobsen flew to Washington that same day and checked into the Madison Hotel at 9:45 P.M.[26]

The next morning, Jacobsen met with Connally for an hour in Connally's office at the Treasury Deparment. *According to Jacobsen, he gave Connally $5,000 in five separately wrapped bundles of $100 bills and said to Connally, "This is part of what we talked about ... there's more where this came from." Thereafter, Connally took the money, walked into an adjoining bathroom for a few moments and then emerged without any sign of the money (Connally denied all of this).*[27]

After leaving Connally's office at 11:15 A.M., Jacobsen stopped by the office of Bill Camp, Controller of the Currency, leaving there at 11:30. From the Treasury Department Jacobsen went straight over to the nearby 15th and M Street office of the American Security and Trust Company, where he opened a safety deposit box, No. 546. After opening the box, he had access to it at 11:43 A.M. and left at 11:47 A.M. *According to Jacobsen, he put the remaining $5,000 from Lilly in American Security and Trust Company's safety deposit box 546.* Jacobsen then returned to the Madison Hotel, where he was staying, and checked out at 2:04 P.M. that afternoon.[28]

Jacobsen and Connally met again on the afternoon of September 23 in Connally's office for almost an hour. It was an appointment which Jacobsen had arranged in advance because Connally's appointment calendar showed a meeting scheduled for 2:30 on that afternoon. Meetings which Connally actually had with people who saw him—rather than scheduled appointments which did not always materialize—are reflected in the Treasury logs. Both Connally's appointment book and the Treasury logs confirmed the meeting with Jacobsen on September 23. Jacobsen, who had checked in earlier that day at 1:00 P.M. at the Madison Hotel,

*claimed that during his meeting with Connally, he asked Connally
if he was ready for the rest of the money. Connally was. Jacobsen said he would bring the money the following day, and they
made arrangements for him to do so the next morning (Connally
denied this).*[29]

The next morning was a busy one for John Connally. He had a
cabinet meeting at the White House at 9:00. From 9:30 to 10:30,
he met with Nixon and Federal Reserve Board chairman, Arthur
Burns. At 11:00, he was to appear on Capitol Hill and testify before the Senate Foreign Relations Committee. Nevertheless, Connally found time to squeeze in an unscheduled meeting with Jake
Jacobsen that morning at 10:35. In other words, instead of going
straight from the White House at 10:30 to Capitol Hill to testify
before the Senate Foreign Relations Committee at 11:00, Connally
stopped off at his office to meet with his old friend Jake Jacobsen.
The meeting was *not* listed on Connally's appointment calendar.
That same morning, at 9:20 A.M., Jacobsen had visited the 15th
and M Street branch of the American Security and Trust Company, where he had access to safety deposit box 546 for the last
time. *According to Jacobsen, he had withdrawn from box 546
the remaining $5,000 he had received from Lilly and brought it to
his meeting with Connally at 10:35. When Connally arrived after
his meeting with Arthur Burns and Nixon, Jacobsen gave the
$5,000 to Connally, who again went into the bathroom and
emerged from it without any sign of the cash (Connally denied
this).*[30]

Despite Connally's denial, the fact remains that Jake Jacobsen
was important enough for Connally to squeeze in an unscheduled
ten-minute meeting between weighty state duties. As the Watergate Special Prosecution Force observed in its internal memorandum on this incident:

> What is significant for these purposes, however, is Connally's
> inability to explain what the brief ten-minute meeting was about on
> September 24, after the date on his log was explained to him. . .[31]

Why had Jacobsen entered his safety deposit box on the morning of September 24, 1971, and then walked over to the Treasury
Department to see Connally for a brief ten minutes? *According to*

Connally, Jacobsen had told him the day before that the Home Loan Bank board wanted Jacobsen out of the savings and loan business and was trying to force him to sell under an unfair formula. Jacobsen asked Connally to raise the question with Thruston Morton, the Chairman of the Home Loan Bank board. Connally subsequently made two phone calls to Thruston Morton on September 23. As for September 24, according to Connally, *"I think Jake came by to find out if I had learned anything."*[32] The prosecutors were skeptical. Came by? To see if he had learned anything? This takes a personal meeting? What are telephones for, the government was to ask subsequently at Connally's trial.

After that meeting, John Connally had few formal contacts with the milk producers and Jake Jacobsen, except for one meeting in March of 1972, with Jacobsen and the new AMPI general manager, George Mehren. They talked about the IRS investigation of AMPI's illegal campaign contribution to Lyndon Johnson, and the Justice Department antitrust suit against AMPI. By this time, Jacobsen and Connally were nowhere as involved with each other as they had been during the first few months of Connally's tenure as secretary of the treasury. Indeed, it was the frequency of their meetings during Connally's early tenure as secretary of the treasury which raised eyebrows at the Watergate Special Prosecution Force:

> The closeness of the Jacobsen/Connally relationship can be seen from Connally's Treasury logs. They showed that in the period February 11 to June 30, 1971, the first three and one-half months of Connally's tenure as secretary of the treasury, Jacobsen met with him on ten separate occasions, more meetings than with any nongovernment person; indeed, more than twice as many meetings as with any of Connally's other nongovernment callers. The logs further reveal that these meetings were not of a fleeting nature. The ten meetings last for over seven hours which is more than twice as much time as Connally spent with any other nongovernmental person in his office during the three and one-half months.[33]

Connally and Jacobsen were not to spend that much time together until the fall of 1973, a time which the Watergate Special Prosecution Force labeled "the obstruction period."

By October 1973, the Senate Watergate Committee's investi-

gation of illegal campaign financing during the 1972 election was
in full swing. Those dairy lobby officials who had been swagger-
ing around town with their briefcases full of money from 1970
through 1972 were scurrying for cover, and Bob Lilly was one of
them.

Harold Nelson called Jake Jacobsen on October 24, 1973, with
some bad news. Bob Lilly was in Washington prepared to tell
Senate Watergate Committee investigators about the $10,000 in
cash which Lilly had given to Jacobsen for what Lilly believed was
John Connally's use. Jacobsen promptly got on the phone with
Connally. *According to Jacobsen, he told Connally that the investigation in Washington was zeroing in on money which Jacobsen
had gotten for Connally. Connally then asserted that Jacobsen
never had given Connally any money, and Jacobsen agreed. Connally then entered into a lengthy monologue in which he told
Jacobsen how he was being investigated all over the place.*
Connally was speaking from his Houston law office, by this time
having resigned from the Nixon cabinet, and he told Jacobsen
that he would be in Austin on Friday. Jacobsen agreed to meet
him there at the Sheraton Crest Inn.[34]

Jacobsen arrived at Connally's suite at the Sheraton Crest at
9:00 A.M. Connally's wife, Nellie, was with him, and he ordered
coffee from room service for the three of them. Sam Barnett, a
waiter who had known Connally over the years, brought the coffee.
They talked. *Jacobsen told Connally that he was not going to tell
anyone that he had given Connally the money. Connally said fine.
They then discussed how to handle what Lilly was telling the
Senate Watergate investigators. Connally suggested that he give
$10,000 back to Jacobsen, who would put it in his safety deposit
box and say it was there all the time. But if they did that, how
would they explain Jacobsen's retaining the money for two and a
half years? Connally had an answer. Jacobsen could say that he
offered it to Connally to give to other candidates when Connally
was secretary of the treasury but that Connally refused to accept
it because of his anomalous position as a Democrat in a Republican administration: as a Democrat in a Republican administration
he did not want to give it to Democrats, and as a member of the
Democratic party he could not give it to Republicans. Better still,*

*suggested Connally, Jacobsen could also say that a year later he
had reoffered Connally the money, after Connally became chairman of Democrats for Nixon, and that Connally declined the
money then as well because milk producers were being publicly
criticized for throwing cash around and Connally did not want to
accept large cash contributions as a matter of policy. As for any
further delay, Connally suggested, all the controversy surrounding
Watergate could explain why the money had not yet been returned
to Lilly (Connally denied it happened this way).*[35]

The meeting ended around 10:00 A.M., and Jacobsen returned
to his law office. Shortly after 2:20 that afternoon, Jacobsen was
subpoenaed to appear before the federal grand jury. Jacobsen
promptly called Connally back at the Sheraton Crest, but Connally
was not there. Jacobsen left a message asking Connally to call
him back.[36]

On Sunday, October 28, back home in Houston, Connally called
Jacobsen. *According to Jacobsen, he told Connally that he had
been subpoenaed to appear before the grand jury, and Connally
said "Jake, we've got to get that money back in your hands right
away," and told him to charter an airplane from Austin to Houston and be in Connally's law office on Monday morning, October
29, to pick up $10,000. Connally has denied saying this and has
denied that Jacobsen told him over the phone on October 28 that
he had been subpoenaed to appear before the grand jury. According to Connally, Jacobsen had already told him about being subpoenaed by the grand jury. Connally claimed that on Friday, October 26, when he returned to his hotel room at approximately
5:00 P.M., he found a message to call Jacobsen. Connelly then
called Jacobsen, who said that he needed to talk to Connally.
Connally told him to come right over, and he was there by 5:20
P.M. They talked for approximately fifteen to twenty minutes.
At the meeting, Jacobsen told Connally of his subpoena and the
involvement of Connally's name. Connally asked Jacobsen why
his name was to be involved, and Jacobsen explained that Lilly had
mentioned Connally's name in connection with conversations
Jacobsen had regarding possible contributions. Jacobsen then told
Connally, "I am prepared to testify that our conversations never
took place and I never heard of any $10,000." Connally told*

Jacobsen he could not do that because the conversations did take place, and they would have to be testified to. While Connally was not happy about this, it was "just one of those things."[37]

Hotel records from the Sheraton Crest did not support Connally's version. The records of Connally's stay at the Sheraton Crest showed *no* local calls. All local calls at the Sheraton Crest are posted to a guest's record by the night auditor. Calls made between a night-auditor posting and checkout time are added at checkout time. The procedure is automatic, and customers are charged for all calls, local and long distance. According to these automatic records, therefore, Connally did not return Jacobsen's call at 5:00 P.M., as he claimed.[38]

On Monday morning, October 29, Jacobsen arranged to charter a private plane. He then called Connally's law office and talked to Connally's secretary to arrange an 11:00 appointment that morning. After the call, Jacobsen drove out to the airport and flew to Houston, arriving around 10:30 A.M. He promptly went to Connally's office, arriving shortly before 11 A.M. *According to Jacobsen, he and Connally again went over the same story they had concocted two days earlier at the Sheraton Crest Inn. After that, they decided they should have a good reason why Jacobsen was seeing Connally that day. Connally suggested that the subject of the meeting be the discussion of the delay in the processing of a bank application by a Connally client named Gus Wortham. Connally then left the room and returned ten minutes later with a cigar box. The box contained $10,000 in small bills and "a rubber glove or rubber gloves." Connally threw the glove or gloves into the waste basket and handed the money to Jacobsen with the comment that the money should be all right because it was "old enough."*[39] Jacobsen then left Connally's office and went to the airport. His charter plane from Ragsdale Aviation took off at 12:35 P.M. and arrived in Austin at 1:30. Jacobsen promptly went to the Citizens National Bank in Austin and entered safe deposit box 865 at 2:00 P.M. *According to Jacobsen, he placed in the box $10,000 in small bills he had received from Connally.*

Connally denied all this. According to Connally, he called Jacobsen on Sunday, October 28, because he was taking care of a

*number of last minute matters before leaving for a two week
trip on October 29. It occurred to Connally that he had done
nothing about a request from Gus Wortham, one of his firm's
oldest and most important clients, to see whether anything unusual
was delaying a bank charter application. Wortham had made the
request on October 10, 1973. Because of Spiro Agnew's resignation on October 10 and all of the calls which followed, Connally had forgotten about Wortham's request and did not think of
it again until October 28. When he thought of it he promptly
called Jacobsen and asked him to come over the next day and
charter a plane, if necessary.*[40]

So, October 29 was a "very, very busy day" for Connally. The
prosecutors wondered why, if it was so busy, Connally did not
handle the matter with Jacobsen over the phone. Connally's reply:
"Well, yes, I think the reason I did not is it would have taken me
longer, I think, on the telephone than it would have taken in
person. . . ."[41] *According to Connally, Jacobsen got to his office
around 10:45 in the morning and they spent twenty minutes together, during which Connally described Wortham's bank application matter, which he wanted Jacobsen to look into. Jacobsen
was then given two $100 bills by Connally to pay for the charter
flight.* Connally admitted, however, that he did not give Jacobsen
any files or folders in connection with the bank application. Connally's client, Gus Wortham, confirms that he had initially discussed with Connally the question of any unusually caused delay
in his bank charter application at a meeting in Connally's law
office and had mentioned the matter again to Connally on October
10, 1973. As late as April, 1974, Connally had yet to provide an
answer for Wortham.

*Connally also claimed that one of the reasons for having Jacob-
sen charter a plane and fly to Houston on Monday morning,
October 29, was that he had observed that Jacobsen was depressed, and "I asked him to come to Houston October 29, so I
could personally look at him and impress upon him the seriousness and interest I had in this bank charter and to judge for myself whether or not he was going to be aggressive enough and
outgoing enough."* But wait a second, the prosecutors asked, had
Connally not seen Jacobsen just the Friday before on October

26? Well, yes, he had seen him that previous Friday but that meeting "was an extraordinary meeting in terms of the subject matter. . . . I did not pay any attention to what his demeanor and what his attitude was or anything else."[42]

But why have Jacobsen fly to Houston just to talk about this matter? Connally:

> *"I told Jake I was leaving for Europe . . . and asked him to fly to Houston. . . . He said he might not be able to get a timely commercial flight, so I suggested he charter an airplane. When he arrived . . . I explained the Gus Wortham problem . . . and asked him to determine . . . what might be causing the delay."*[43]

Did Connally give Jacobsen any files or folders in the Wortham matter? *"No, I only had a few notes jotted down. . . ."* Did Connally pay for the charter flight? "Yes, I gave Mr. Jacobsen two $100 bills. . . ."[44]

On the next Friday, November 2, 1973, Jacobsen flew to Washington and testified before the Watergate grand jury. That Saturday he was interviewed by staff members of the Senate Watergate Committee. The following Wednesday, November 7, 1973, his deposition was taken in Ralph Nader's lawsuit over the March, 1971 milk price support increase. Before the grand jury, Jacobsen told the coverup story which he claimed he and Connally had agreed to. When asked if he would permit federal agents to look into the Austin safety deposit box to see whether the $10,000 was still there as he had claimed, Jacobsen agreed. Jacobsen then returned to Austin and was contacted by attorney Larry Temple, who suggested that he come to Temple's office for a meeting with Marvin Collie, Connally's law partner and the "best tax lawyer in Texas," on Friday, November 9. Jake met Collie at Temple's office on that date and was "completely debriefed." He told Collie, among other things, about the grand jury questions concerning the $10,000, his answers, and his consent to an inventory of the money. Jacobsen was told that Connally was due back in Texas on Monday, November 12, and would testify himself before the grand jury later in the week. *Connally denied setting up the debriefing and claimed that he did not know what Jacobsen said at the debriefing. Was Connally ever informed of what Jacobsen said before the grand jury? "No" responded Con-*

nally. "*As a matter of fact, I got back on the twelfth, as I recall, and appeared on the fourteenth, and I really don't know to this good day what Mr. Jacobsen [said] when he appeared before the . . . grand jury.*" Nevertheless, Jacobsen did place a two-minute person-to-person long distance phone call to Connally's home in Houston on Monday, November 12, 1973, the day Connally was due to return home.[45]

On November 14, 1973, Connally testified before the Watergate grand jury. His story corresponded to what Jacobsen said they had agreed to on October 26. But in answering the prosecutor's questions, Connally was careless or forgetful, possibly both. Connally testified that Jacobsen had offered him $10,000 to give to any candidate of his choice. Connally said he declined because as a Democrat in a Republican administration he did not want to be involved. The following exchange then occurred:

Question: Did Mr. Jacobsen tell you the source of this $10,000?
Connally: No, he did not.
Question: He didn't tell you it had come from AMPI or the dairy industry?
Connally: No.[46]

These were crisp, clear, precise answers. Connally was not worried, not then. When he next testified before the Grand Jury (on April 11, 1974, after Jacobsen himself had been indicted for perjury) Connally changed his story.

Connally: My best recollection is that he said that the milk producers were going to start making some political contributions in 1971 and that they would be contributing to candidates in both parties and that there was $10,000 available then to be given to any committee or candidate or campaign that I would designate.
Question: So it is correct to say that when he first mentioned the $10,000 to you, he did mention the source of the money?
Connally: Yes. I don't recall that he mentioned anything in particular. As I recall, he did say milk producers.[47]

On Saturday, November 24, 1973, George Christian, former press secretary to both Connally and Lyndon Johnson, called Jacobsen and asked him to meet Connally at Christian's home in Austin, Texas, the next day. Jacobsen agreed. Connally, professing to be worried about news stories that he was being investigated by

the Watergate prosecutors, had telephoned Christian, asking him
to get together Jake Jacobsen and Larry Temple to discuss
strategy, i.e., should Connally demand a public hearing before
the Senate Watergate Committee or should he choose another forum
and, if so, what?[48]

Christian, his wife, and some of their children were there when
Jacobsen arrived. Connally was not. Mrs. Christian left to play
tennis. Connally arrived, carrying an attache case. The three men
talked briefly after Connally arrived. When Christian's four-year-
old woke up, Christian left Connally and Jacobsen to go upstairs to
dress him, take him to the kitchen, and fix a bowl of cereal. By
the time Christian returned, Connally and Jacobsen were ready
to leave.

What happened in the short time Christian left Connally and
Jacobsen alone? *According to Jacobsen, Connally told him some
of the money in the first batch he gave Jacobsen on October 29
contained "Schultz bills" (money issued while George Schultz,
Connally's successor, was secretary of the treasury) and hence
some of the bills were issued later than they should have been to
be sitting in Jacobsen's safety deposit box all that time. So Connally had brought Jacobsen another $10,000 to replace the Octo-
ber 29 batch of money. But wait, said Jacobsen, he promised the
Watergate prosecutors that they could inspect his safety deposit
box, and if he signs in to that safety deposit box the day before
they're scheduled to have the FBI inventory the contents on Tuesday, November 27, well, he just can't do it. "John, I told them
in Washington I wouldn't touch that . . . box until the FBI inspects
it. I'll have hell getting in." Connally was not concerned. He believed it could be worked out. "Well, Jake . . . you simply got to
get in that box without leaving a trace." Connally opened his brief
case. There was a bundle wrapped in newspaper inside. Connally
then apparently changed his mind and suggested they make the
money exchange outside.* Christian reappeared from the kitchen
as Jacobsen and Connally were leaving. They called goodby to
him. *According to Jacobsen, he and Connally both entered Connally's car. Connally again opened his attache case. This time, the
newspaper wrapped bundle was taken out and passed to Jacobsen.
The newspapers contained a second $10,000 from Connally. Con-*

*nally told Jacobsen to keep the first $10,000 until the matter
blew over. Jacobsen then went home with the money.*[49]
   Connally denied that he passed any money to Jacobsen at Christian's house. Christian backed up Connally's story in a small but
crucial respect. He testified at Connally's trial that he did not see
Jacobsen get in Connally's car. This is important because that is
where, according to Jacobsen, the money exchange took place.[50]
   Connally's overall testimony about the meeting at Christian's
house, however, did not satisfy the prosecutors. *Connally denied
that he and Jacobsen discussed their respective grand jury testimony at Christian's house.* According to Connally, "*I asked how
he was feeling. We engaged in a little small talk.*" And what did
Connally say to Jacobsen as they walked away from Christian's
house to their cars? According to Connally:

> *I asked him if he'd heard anything on the Gus Wortham matter.
> He gave me an indirect answer. I thought to myself that he hadn't
> done anything on it. So I gave him, perhaps, a little unwanted ad-
> vice. I said, "Jake, you're going around with this hangdog look,
> you won't see your friends, you go around looking like a sheep-
> killing dog. You've got to get a hold of yourself. It's no disgrace
> to go broke. Take your bankruptcy, and then go practice law.
> Other people have done it.*"[51]

   On another key point, however, Christian's testimony supported
Jacobsen. Did Connally, in fact, have a briefcase with him. Yes,
he did, recalled Christian. Was it opened in Christian's presence?
No.[52] Well then, the prosecutors wanted to know, what was the
meeting for? To hear Connally tell it, all they did was engage in
small talk, culminating in a dutch-uncle lecture to Jacobsen as they
walked away. This was a strategy meeting which Connally asked
Christian to arrange and yet Christian spent most of the time with
his four-year-old son rather than with Connally and Jacobsen.
Further, Connally had come from only a few blocks away—his
daughter's house. And he returned to his daughter's house. Why
did he bring a briefcase with him to Christian's house, the prosecutors asked, since he never opened it in Christian's presence? For
the prosecutors, Jacobsen's story supplied an answer, i.e., the transfer of money with no witnesses.
   The next day, Jacobsen saw his former law partner, Joe Long,

who was also an official of the Citizens National Bank. He asked
for Long's help in getting into his safety deposit box without a
record being made of it. Long agreed to help. Late in the day
after the bank was closed to customers, Long got the master key,
and the two of them went into the vault together. Long took the
contents of box 865 (which Jacobsen had last entered on October
29, after he had seen Connally) and gave it to Jacobsen. Jacobsen gave Long a package, which Long then placed into Jacobsen's
other safe deposit box at the bank, box 998. As the two men
started to leave, they encountered trouble. No one was at the desk
in front of the vault when they went in. When they left, they found
the desk occupied by Virginia Strong, who was in charge of maintaining records for access to safety deposit boxes. She told Long
to sign his records for access to safety deposit box 555. Long did,
and Strong initialed Long's access records. Jacobsen did not
sign any access records.[53]

Although he did not know it at the time, the next day, November 27, was the beginning of the end for Jacobsen. FBI agents
called him and asked to inventory his safety deposit box. Later
that same day, he met two FBI agents at Citizens National Bank,
where the contents of Jacobsen's box 998 were inventoried. It was
the same box into which Long put a package from Jacobsen the
day before. The package contained $10,000. "I'm sure glad it's
all there," said Jacobsen to the agents.[54]

Not so fast, Mr. Jacobsen. So those bills have been in that box
untouched since May, 1971, have they? Yes. You didn't give the
money to Connally? No. You didn't use it yourself? No. You don't
know of anyone else who ever used the money? No. It's really
been there since May of 1971, when you first put it in? Yes.
Well, now, the inventory shows a total of four $100 bills, ninety-
two $50 bills, and two hundred fifty $20 bills. The $100 bills are
all "old enough"—they come from the 1960s. The fifties and
twenties, however, are another story. And they sink Jacobsen *and*
his story. Because in May, 1971, when Jacobsen claimed he put
the money in the box, no less than twenty of the bills were not yet
in circulation.[55]

On November 27, 1973, however, Jacobsen did not know he
was in danger. He still thought he was safe. Indeed, he went to

Washington on December 14 and testified before the Senate Watergate Committee and repeated his lie about the money being there since May of 1971.

A week before that appearance in Washington, a curious event happened. Attorney Larry Temple, a friend of both Jacobsen and Connally, called Jacobsen and said that his secretary was bringing an envelope over to Jacobsen. The envelope had the return address of John Connally's law firm. Inside were all sorts of things which were helpful to Jacobsen in preparing for his appearance before the Senate Watergate Committee: a transcript of Connally's appearance before the Watergate Special Prosecution Force, a transcript of Connally's appearance before the Senate Watergate Committee, and a very detailed digest of Connally's grand jury testimony. Jacobsen avidly read the contents. Temple was so helpful because, according to the prosecutors, "he picked the envelope up at Connally's law firm on Friday, December 7, 1973, with instructions to make it available to Jacobsen." Temple glanced at the contents of the envelope but did not read it, did not make copies, and did not discuss it with anyone. He simply had his secretary deliver it to Jacobsen the following Monday, December 10.[56]

Why did Connally arrange to have all these helpful documents given to Temple and Jacobsen? *According to Connally, he told his secretary to get a transcript of his testimony before the Senate Watergate Committee to Larry Temple. Connally claimed he was unaware whether the transcript of his appearance before the Watergate Special Prosecution Force was included. In a talk with Temple, Connally said that he told Temple he was going to send the material to him because, if the Senate Watergate Committee reconvened, he wanted the opinion of Christian and Temple on his testimony. Connally admitted, however, that he told Temple he could show the Senate Watergate Committee transcript to Jacobsen.* Temple, however, did not study the transcript and did not make copies; he merely glanced at them and promptly delivered them to Jacobsen. Connally made no copies for anyone else besides Jacobsen.[57]

Even though he was prominently mentioned and probably

would have been appointed as Spiro Agnew's successor, but for the Watergate-weakened condition of Richard Nixon, 1973 had not been a good year for John Connally because of his Senate Watergate Committee and Grand Jury appearances. 1974 did not promise to be any better and it was not.

On February 21, 1974, Jacobsen was indicted by a federal grand jury in the District of Columbia for perjury by having testified before it that he had not touched the $10,000 in safety deposit box 998 between May, 1971 and November, 1973.[58] The grand jury indictment also specified that Jacobsen had gotten money in 1971 from AMPI "on the representation that such money was to be paid to a public official for his assistance in connection with the price support decision." The public official, of course, was John Connally, although he was not publicly named at that time.

On March 16, 1974, Jacobsen pleaded innocent to the charge of lying to the federal grand jury. While newspaper reports in the *Washington Press* speculated that Jacobsen's indictment was designed by the Watergate Special Prosecution Force to pressure him into cooperating with the government, Jacobsen's lawyer, Charles A. McNellis, denied this and told reporters that Jacobsen was not plea bargaining with the special prosecutors and did not contemplate doing so.[59]

On April 10, 1974, the leaks started. Washington columnist Jack Anderson reported that the Watergate prosecutors were investigating bribery charges against John Connally. According to Anderson:

> The crack FBI squad, which is assigned to the special prosecutors' office, has dug up evidence that Connally pocketed $10,000 from the Associated Milk Products [sic], Inc., and hastily returned the cash after the dairy lobby came under investigation. . . . Our FBI sources say the alleged $10,000 bribe was passed by AMPI official Robert Lilly to lobbyist Jake Jacobsen, who delivered it to Connally.[60]

Anderson's column went on to report the government's version of the two occasions on which Connally allegedly returned money to Jacobsen. The column then inaccurately referred to the milk price support decision of 1971 as occurring in March 1972, on the day after Connally held a meeting with Jacobsen and AMPI officials Mehren and Nelson about a government antitrust case against AMPI. Faced with these leaks, Connally responded to reporters: "I have categorically denied I received the money, and I do so today." Jacobsen declined to comment.[61]

Anderson kept milking his story. On April 15, 1974, another Anderson column reported that the Watergate special prosecutors had asked the White House for tapes of conversations between Nixon and Connally on milk matters. Anderson noted that the prosecutors were particularly interested in Nixon-Connally conversations prior to the milk price support decision—although Anderson once again made the mistake of placing the matter in 1972 instead of 1971. Anderson also reported that the prosecutors were interested in Nixon's conversations with former Attorney General John Mitchell on the milk case.

By April 20, 1974, further leaks hit the news. Press reports indicated that:

> Texas lawyer Jake Jacobsen has sent word to Watergate prosecutors that he is prepared to testify that former Treasury Secretary John B. Connally took $10,000 for helping a giant dairy cooperative, according to an informed source.[62]

Jacobsen, when reached at his home, refused to comment: "That is what my lawyer instructed me to do." Jake's lawyer, Charles McNellis, wasn't talking either. In sharp contrast to his statements to the press in March, 1974, that no plea bargaining was under way or contemplated, McNellis refused to confirm or deny the published reports of his client's plea bargaining.

When the Jacobsen plea bargaining story broke on Saturday, April 20, 1974, John Connally was not immediately available for comment. He was on his way to Bangor, Maine to deliver the keynote address to eighteen hundred delegates at a state Republican convention. When the story broke, there were a few rats seeking to leave what looked like a sinking ship. "The last thing we need is for a former Democratic governor of Texas to come up here and give the keynote address and then get indicted," entoned Harrison L. Richardson, an otherwise obscure state senator running for governor, who was quoted for no apparent reason other than that he had something nasty to say about John Connally.[63]

Connally, on the other hand, held up well at an impromptu press conference at the airport: "There is nothing new—the story is not changed." It was, said Connally, "a typical offer of a campaign contribution by a friend which I did not accept."[64] Connally then went on to deliver his keynote address at the state convention, again asserting his innocence, and praised the embattled Nixon for ending the Vietnam War and improving international relations with China and the Soviet Union.

Nevertheless, the damage was done. Nationally syndicated columnist George Will in the next issue of William Buckley's *National Review* devoted a column to John Connally's predicament captioned "That's All for A While," in which he suggested that, whatever the truth about the leaks on Connally then surfacing, Connally as a presidential candidate in 1976 was no longer "viable."[65]

The day before his trip to Maine, Connally met for ninety minutes in the White House with Richard Nixon. It is not known what they discussed, and Connally professes today not to recall, but Connally has denied ever discussing with Nixon his case or Jacobsen's plea bargaining with the Justice Department.

Connally's denial should be taken with a grain of salt, however, given the timing of Connally's meeting with Nixon—nine days after Jack Anderson first carried leaks about Jacobsen's implication of Connally and the day before the appearance of press reports that Jacobsen had agreed to a plea bargain. Even more significant is that eleven days later, Nixon himself called Deputy Attorney General Laurence Silberman to talk about the Connally case. Nixon had his back to the wall with his own impeachment proceedings yet displayed surprising familiarity with precise details of Connally's case. He told Silberman that the Justice Department's arrangement to drop federal charges pending against Jacobsen in Texas was an abuse of the plea bargaining process. When Silberman told him the Connally case was being handled by the Watergate Special Prosecution Force and not the Justice Department, Nixon was well-enough informed to reply that the original charges against Jacobsen were brought by the U.S. Attorney in Abilene, Texas, who was under the jurisdiction of the Justice Department. Nixon ordered Silberman to check into the matter and report back to him within an hour. The federal charges pending against Jacobsen in Abilene involved allegations that Jacobsen had misapplied bank funds. Silberman learned this when he called Assistant Attorney General Henry Peterson promptly after receiving the call from Nixon. Peterson told Silberman that Jacobsen had agreed to plead guilty in the Connally case to one felony count of paying the illegal $10,000 gratuity to Connally. Peterson said that under these circumstances, there was nothing unusual or improper about dropping charges in one jurisdiction in exchange for a guilty plea and cooperation in another case in a different jurisdiction. Silberman called Nixon back within the hour and told him what he had learned from Peterson. Nixon was not satisfied. Highly agitated, Nixon demanded that the plea bargaining with Jacobsen be terminated and that the Abilene charges be prosecuted. Silberman told him that Henry Peterson would not agree to this. Nixon was furious: "Well, Henry Peterson will just have to go." Nixon told Silberman to report back to Alexander Haig. After conferring with his aides, and deciding that he would have to resign if Nixon didn't withdraw his order on the Jacobsen plea bargaining, Silberman called Attorney General William Saxbe, who was off pheasant hunting. Saxbe, a remarkably plainspoken Ohio politician who had served as his state's Attorney General before being elected to the United States Senate in 1968, listened to Silberman's story and told him to forget about resigning: "You're out of your mind. I am not going to resign and neither should you. They can fire us if they want." Quick to make decisions, Saxbe gave Silberman his marching orders: "Tell the White House to piss up a rope. And don't bother me again; the hunting is good." Silberman promptly called Alexander Haig at the White House and advised him that he and Saxbe would not obey Nixon's orders. Haig professed ignorance of Nixon's earlier call to Silberman and said he would get back to Silberman on it. Some semblance of sanity was apparently restored at the White House as Haig called Silberman back later that day and told him to "just forget the whole matter."[66]

As a result, seven weeks later, on June 21, 1974, Associated Press reporter Brooks Jackson, who had been covering the milk fund scandal almost from its inception, reported that Jacobsen's lawyers had worked out a plea bargaining agreement with the

Watergate prosecutors to let Jacobsen testify against John Connally in exchange for pleading guilty to a reduced felony charge of giving a gratuity to a federal official.[67]

On July 29, 1974, both Jacobsen and Connally were indicted. Jacobsen was charged with giving a gratuity to a federal official. Jacobsen's plea bargaining deal was consummated less than two weeks later on August 8, 1974, before Chief U.S. District Judge George L. Hart, Jr., in Washington when he pleaded guilty. Jacobsen appeared subdued before Judge Hart. His voice was low, almost a whisper: "I plead guilty, your honor." Hart asked him if he had actually made the payment to Connally as alleged in the indictment? "Yes, sir," was Jacobsen's whispered reply.[68]

For his alleged role, Connally was indicted on seven separate counts. The first two counts against Connally charged him with acceptance of an illegal gratuity, once on May 14, 1971, and again on September 24, 1971, for performing an official act—recommending a price support increase to the president.

These are the only two counts on which Connally was to subsequently stand trial. The other five counts all dealt with the cover-up, which the prosecutors believed Connally and Jacobsen had contrived. At Connally's request, the court ordered these counts to be tried separately but they never were. They were dropped after Connally's acquittal. The third count charged Connally with conspiring with Jacobsen to obstruct justice and make false declarations before the Senate Watergate Committee and the grand jury. The fourth count charged Connally with perjury before the same two government bodies. The fifth and sixth counts charged Connally with giving false testimony before the grand jury on November 14, 1974, when he testified that he had not spoken to Jacobsen for a "long time" about the $10,000 and that his "only" contact with Jacobsen in the three to four weeks preceding his November 14 appearance was at Connally's law office in Houston on October 29. The prosecutor believed that Connally was attempting in his grand jury appearance to cover up his October 26 meeting with Jacobsen at the Sheraton, where they did in fact discuss the $10,000. During his initial November grand jury appearance, Jacobsen had told a similar story, which did not reveal his October 26 meeting with Connally or their discussion of the $10,000. The counts were technically accurate, for in his second grand jury appearance on April 11, 1974, Connally changed his story and admitted that he did have a meeting with Jacobsen on October 26 and that they did discuss the $10,000.

The final count charged Connally with perjury for denying that Jacobsen had told him in their phone conversation on Sunday, October 28, about the grand jury subpoena Jacobsen had received the previous Friday afternoon, following their early morning meeting at the Sheraton. This was the first opportunity Jacobsen had had to talk with Connally since he received the subpoena, and the prosecutors believed it incredible that Jacobsen would not have told Connally about it, particularly since the two had discussed the general problem of the $10,000 only two days earlier. The prosecutors believed Connally lied about the October 28 phone call to cover up what they thought was the true purpose for flying Jacobsen to Houston on a chartered plane the next day —Connally's returning $10,000 to Jacobsen.

Why did the Watergate Special Prosecution Force seek an indictment of John Connally? What was behind it? What did they think the weaknesses of their case were? The strengths? Did they weigh the effects on Connally's possible presidential candidacy?

Government prosecutors do not often answer questions like these, and the government was not eager to have these questions answered. It took a Freedom of Information Act lawsuit by William Dobrovir and the Fund for Constitutional Government to pry the information loose—all of the thirteen prosecution memoranda of the Watergate Special Prosecution Force relating to the milk fund investigation, among which was the June 18, 1974, prosecution memorandum on John B. Connally.

That memo reveals that Watergate prosecutors perceived three major weaknesses in their case against Connally. First, Jacobsen was pleading guilty to a reduced felony charge of attempting to give a gratuity to a public official, and other more serious charges against him were being dropped in exchange for his testimony against Connally. Second, it was hard to believe that Connally would violate the law for a mere $10,000. Third, Jacobsen could not account for an additional $5,000 he had received from Bob Lilly over and above the initial $10,000.

The prosecutors were not concerned about the first weakness —Jacobsen's being let off easy on other crimes in exchange for testifying against Connally. "The answers to this were the usual answers since this problem is certainly not unique," responded Prosecutors Frank Tuerkheimer and John Sale in a memorandum to Henry Ruth. Further, they believed there were additional points to balance this weakness, specifically, the fact that Jacobsen did not attempt in the words of the prosecutors to "give us" other persons whom Jacobsen knew the special prosecutors were interested in—Harold Nelson and George Mehren of AMPI. If Jacobsen were fabricating his story, the prosecutors reasoned, "it would only be logical for him to start with the story that might allow him to try to salvage his position in Texas and his law license and liberty." Further, the prosecutors believed that Jacobsen had ample opportunity to embellish his story against Connally, which he did not take. Jacobsen could have said that Connally insisted upon money before the second price support decision, or he could have told them that he had talked to Connally about the possibility of personal cash for Connally before both price support decisions. All of this would have strengthened the prosecution's case. Jacobsen, a lawyer himself, obviously knew this and also knew there would be no way to disprove his story. It would have been Jacobsen's word against Connally's. The circumstantial evidence on the delivery of the money would be the same. And there is no question that Jacobsen had sufficient opportunity before the price support decision to see Connally alone to make such a story convincing. As the prosecutors state in their memorandum, "Certainly Jacobsen was alone enough with Connally in early 1971 to come up with a more dramatic and revealing story" than he actually did.[69]

The second major weakness—that it was hard to believe that Connally would violate the law for a mere $10,000—was discounted by the prosecutors for a number of reasons. For one thing, they considered it unlikely that in 1971 Connally really believed he was risking anything by dealing with a person like Jake Jacobsen, whom he had known for twenty-five years. Further, Jacobsen himself was dealing with Bob Lilly, who was merely a professional acquaintance, an employee of a major client, and hence much less close to Jacobsen than Jacobsen was to Connally. It was Jacobsen, therefore, who was taking the real risk, not Connally. Accordingly, as the prosecutors state in their memorandum, "It is inconceivable that Jacobsen should in his April 28 conversation with Lilly distort the truth in favor of criminality by asking for money for Connally personally when it was supposed to be for political candidates." Asking Lilly for the money for Connally's personal use was Jacobsen's *admission against interest* to Lilly that Jacobsen was going to engage in a criminal act, i.e. paying a gratuity to a cabinet officer for the price support decision. If that were not true, why did Jacobsen not just tell Lilly he wanted the money so that Connally could make campaign contributions? Finally, the prosecutors suggest that with AMPI talking about a commitment of $2 million to the Nixon campaign in the spring of 1971, and John Connally's detailed knowledge as evidenced in the March 23 meeting with Nixon and his advisers about the money being raised for the dairy cooperatives, "it would be almost an insult to Connally to give him a mere $10,000 for campaign contributions when AMPI was talking in terms of $2 million. Unlikely as it is that Connally would violate the law for a mere $10,000, it is equally unlikely that AMPI would lawfully deal in such small sums with him." In addition, the prosecutors pointed out in the memo that in August of 1972, when Connally claimed that Jacobsen offered him the money for a second time (which Connally asserts he declined), neither Jacobsen nor his law firm were still serving as attorneys for AMPI. Their relationship had been severed by George Mehren, AMPI's new general manager.[70] Why would Jacobsen still offer AMPI money to Connally after he had been fired as its lawyer?

The third major weakness—the third $5,000 from Bob Lilly—was perceived by the prosecutors as the most important. The story behind this third $5,000 is crucial because Edward Bennett Williams subsequently used it at the trial to destroy Jacobsen's credibility and effectively impress on the jury's minds the possibility that Jacobsen himself had taken the entire $15,000 and given none of it to Connally.

Here is that story. On October 13, 1971, Bob Lilly boarded Braniff flight 415 in San Antonio enroute to Washington with a layover at Dallas. While in Dallas, Lilly received a call from Jacobsen, who had earlier called him in San Antonio and learned he was on his way to Washington. Jacobsen told Lilly that he was soon going to Washington and wanted to be able to tell Connally that AMPI would have another $5,000 (the "third $5,000") for him in cash and that the money would be in "Jake's safety deposit box at the Citizens National Bank in Austin in a short time." "Okay," said Lilly, "you can tell Mr. Connally that." The next day, October 14, 1971, Jacobsen met for a half hour with Connally in his office in the Treasury Department. On November 3, Lilly received a $5,000 check payable in cash from Stuart Russell. On the morning of November 10, 1971, Lilly flew from San Antonio to Austin. Upon arriving, Lilly walked from the plane into the airport building and noticed a meeting already in progress in the airport coffee shop between Dave Parr and Tom Thompson of AMPI, and Jake Jacobsen. They all saw Lilly and motioned him over. Lilly joined the meeting, which lasted for over an hour. Jacobsen's law partner, Joe Long, subsequently joined the meeting. At one point in the meeting, Jacobsen handed an envelope to Dave Parr. "This is $5,000 for Mills," said Jacobsen. No money was counted and Parr pocketed the envelope. Lilly, after the meeting, proceeded directly to the Citizens National Bank, where he cashed Stuart Russell's check. He then went up to Jacobsen's law office and, in the presence of Jacobsen's partner, Joe Long, delivered the third $5,000 to Jacobsen. It was still before noon, and Jacobsen promptly left to visit his safety deposit box. And, sure enough, the records for Jacobsen's safety deposit box 998 showed that he had made an entry at 12:30 P.M., on November 10, 1971, the first time he had entered that box since February 3, 1970. Jacobsen next entered box 998 on December 14, 1971, at 2:20 P.M. and flew to Washington a day later.[71]

The prosecutors' problem was that Jacobsen initially denied any recollection of receiving an additional $5,000 from Lilly. This was an unfortunate memory lapse and, after intensive drilling and confrontation with the damning safety deposit box records, Jacobsen agreed that he must have gotten the third $5,000 from Lilly, "and that if he asked Lilly for the money for Connally, then he must have given it to Connally." Jacobsen told the prosecutors, however, that "he had no recollection of giving it to Connally." This curious lapse of memory coupled with the weak conclusion that he "must have given it to Connally" was used with devastating effect by Edward Bennett Williams in his attack on Jacobsen's credibility at the trial. Prior to indictment, however, the prosecutors rationalized this possibility in their memorandum as follows:

> Connally has little room in which to argue that Jacobsen kept the $5,000 for himself. This is so because at the time he obtained the $5,000, Jacobsen was still a multimillionaire with no storm clouds on the horizon. Since Connally has acknowledged Jacobsen offered him $10,000 in August 1972, two months after Jacobsen was found bankrupt, it makes little sense to argue that he would steal $5,000 while still a very wealthy man.
>
> What the defense can argue is that it makes no sense to cover up for $10,000 when $15,000 was paid. Our answer is that Jacobsen simply forgot about the third payment, and it was only after he went over his safe deposit records (and was told about Lilly's testimony) that he remembered to the extent that he does.[72]

That really wasn't a good answer. It genuinely did make no sense to cover up for $10,000 if in fact Connally received $15,000. Connally and Jacobsen *both* aren't going to forget about the additional $5,000. If Jacobsen got $15,000 from Lilly and told Lilly it was all for Connally, and Connally really received his $15,000, then Jacobsen's story that he and Connally plotted to cover up and replace $10,000 is questionable. Jacobsen's story hinges on his being able to show two separate packages of $10,000 each in his safety deposit boxes, packages placed there, he claims, *after* receiving them from Connally as part of their coverup. But if Connally really received $15,000 from Jacobsen, why didn't he return *$15,000* back instead of $10,000? If Jacobsen was telling the truth about receiving only the two $10,000 payments from

Connally, then it can't be right that he "must have given" the third $5,000 to Connally as well. And if Connally didn't get that extra $5,000, who did? This was the hole in Jacobsen's credibility which Williams was subsequently able to exploit so skillfully.

One possible—and very plausible—explanation for the third $5,000 is that Jacobsen did in fact keep it for himself and did not pass it to Connally. Remember that on the same day Bob Lilly gave Jacobsen the third $5,000 for Connally—November 10, 1971— Jacobsen had just delivered $5,000 in cash to Dave Parr for Wilbur Mills' campaign. Jacobsen also denied billing excess legal fees to AMPI as reimbursement for the Mills contribution. It is not implausible to speculate, therefore, that Jacobsen decided to reimburse himself for the $5,000 he gave to Parr by simply keeping the third $5,000 from Lilly intended for Connally. There is no evidence the prosecutors ever considered this possibility, however, and the third $5,000 remained a major weakness in their case.

Given such a major weakness, which was to lead to the subsequent acquittal of Connally, why did the special prosecutors go forward? Why did they think their case was that strong? Two reasons: John Connally did not tell the same story twice, and the circumstantial evidence backed up Jacobsen's story, not Connally's.

CONNALLY'S CONTRADICTIONS

On November 14, 1973, Connally testified before the grand jury that Jacobsen had not told him the source of the $10,000 being offered nor that it came from AMPI or the dairy industry. On April 11, 1974, Connally had a different recollection and believed that Jacobsen did mention the milk producers.[73]

On November 14, 1973, Connally testified before the grand jury that it had been a long time, perhaps since the fall of 1972, since he had last discussed the $10,000 with Jacobsen. Nevertheless, on April 11, 1974, Connally testified that he had had an "extraordinary conversation" with Jacobsen about the money less than three weeks before the November 14 testimony, i.e., at their meeting on October 26 at the Sheraton Crest in Austin.[74]

Further, on November 14, 1973, Connally had testified before the grand jury that his October 29 business meeting with Jacobsen at Connally's law office in Houston was his *only* contact with Jacobsen in the three to four weeks preceding his testimony. Connally naturally contradicted this testimony as well on April 11, 1974, when he finally admitted the October 26 meeting at the Sheraton Crest.[75]

> The significance of the latter two contradictions is self-evident. There can be little room for doubt that Connally deliberately intended to mislead the grand jury on the nature and frequency of his pretestimony contacts with Jacobsen.[76]

The prosecutors further pointed out that phone records of the thirteen-minute phone conversation between Connally and Jacobsen on October 24, 1973, and their three-minute conversation on November 13, 1973 (one day earlier than Connally appeared before the grand jury), also conflicted with Connally's 1973 testimony before the grand jury.[77]

The circumstantial evidence, however, clinched it for the prosecutors. As they explain in their analysis:

> In a case of this sort, many inferences can be found supporting one witness's version or another. They vary in strength and we do not in this memorandum attempt to point [out] every inference, however weighty or unweighty [which] can be drawn. For example, that all major contacts between Connally and Jacobsen in the events of October 24-29 until Jacobsen's indictment here were initiated by Connally—November 9 debriefing, the George Christian meeting, the transfer of Connally's Washington testimonies— in all cases through *intermediaries*, vaguely supports the notion that Connally was up to something improper. We will not, however, attempt to cull from the facts of this case other arguments of comparable force. Rather, we point to what we believe are the four most significant areas of the circumstantial support for Jacobsen's version.[78]

Those four areas were (1) money, (2) the safe deposit box records in Washington, (3) the safe deposit box records in Texas, and (4) the October 29 meeting in Connally's law office in Houston with Jacobsen.

THE MONEY

The prosecutors were faced with two separate sets of bills: the "October 29" bills which Connally allegedly passed to Jacobsen in Connally's law office, and the "November 25" bills which Connally allegedly gave Jacobsen at George Christian's home. The government recovered these two sets of bills on separate occasions. The "November 25" bills were inventoried by the FBI two days later on November 27. These were supposed to be the "old" safe bills which Connally had given Jacobsen at Christian's house to support Jacobsen's story that the money had sat in his safe deposit box since June of 1971. Somebody erred, however, because sixteen of these bills seized by the FBI on November 27 had not been in circulation on June, 1971, three of them had not been in circulation by October, 1971, and one of them had not been in circulation until March, 1973. It was this discovery which discredited Jacobsen's original story. He claimed he had not touched the bills since he deposited them in June of 1971, and the bills themselves made him out to be a liar. After his cover story proved untenable, Jacobsen then switched to his story about Connally. The bills the government seized on November 27 had, in fact, been given to him two days earlier by John Connally, Jacobsen claimed, but Connally had also given him another $10,000 a month earlier on October 29. Jacobsen offered these bills to the prosecutors.

The prosecutors accepted, and the bills were seized and inventoried on February 28, 1974. The prosecutors considered these "October 29" bills crucial—a litmus test of Jacobsen's credibility. On October 29, 1973, Jacobsen did not yet know he was a target of the Watergate grand jury. It was too early. He had found out only the week before that Lilly was talking to the Senate Watergate Committee investigators. According to Jacobsen, he and Connally were in the midst of their coverup and things were proceeding well. For, if Jacobsen told the truth, *all* of those bills would have been in circulation on October 29, 1973. *If only one of those bills was not in circulation on October 29, 1973* (and keep in mind it is not possible by looking at the face of a bill to know whether it was in circulation on a given date; this takes recourse to Federal Reserve records, which are not routinely available to the general public), then Jacobsen was lying. Only one post-October 29 bill would mean that he had not deposited the money on October 29 as he claimed, but later. Such a lie would have destroyed Jacobsen's credibility for the prosecutors, and they would not have proceeded further. The lie would have broken the symmetry of Jacobsen's visits to Connally being preceded or followed by a trip to Jacobsen's safety deposit boxes.

So what happened? The prosecutors found that of the $10,000 allegedly placed in the deposit box on October 29, 49 of the 280 bills were issued while George Schultz, Connally's successor, was secretary of the treasury. The prosecutors focused on these 49 more recent bills: "Since these bills had been in circulation for such a short period, there is *every chance* that a post-October 29 Schultz bill would appear among the 49." The prosecutors then ascertained through painstaking research that *every one* of the 49 Schultz bills had been in circulation on October 29, 1973, the date when Jacobsen claimed he placed them in his safe deposit box. Despite the fact that over four months had elapsed before the government inventoried the October 29 money, *not a single one* of the 49 Schultz bills had been issued after October 29, 1973. The prosecutors found this convincing: "The absence of such a bill points to an origin of the $10,000 on or before October 29, long before Jacobsen knew of his problems."[79] When Jacobsen offered to take a lie detector test, they declined, considering the total absence of bills circulated after October 29, 1973, to be more convincing than any such test.

THE WASHINGTON SAFE
DEPOSIT BOX RECORDS

The safe deposit box records at the American Security and Trust Company in Washington showed that Jacobsen made his *only* two entries into that box within one hour of meeting with Connally. The prosecutors were persuaded that:

> Since the first entry is made when the box was opened, the only plausible inference is that something was put into the box on that date, May 14, 1971. Since the box was found to be empty after the second entry, it is also compelling that whatever was put into the box on May 14 was taken out on the date of the second entry—September 24, 1971.[80]

Consider this fact in the context of the events of September 23, 1971—Jacobsen's forty-five minute meeting with Connally, which had been scheduled in advance—and those of September 24, 1971, when Connally held a hurried ten-minute meeting with Jacobsen in the midst of an extremely busy day:

> Given their extensive meeting of the day before, their telephone contacts and the hectic day Connally had on September 24, there is considerable inferential support for the idea that the meeting was set up in order for Jacobsen to give something to Connally pursuant to an arrangement of the day before and that in fact he did. This inference becomes all the stronger when taken in conjunction with the emptying of the safety deposit box shortly before the meeting.[81]

TEXAS SAFE DEPOSIT BOX RECORDS

On the same day he visited Connally in Houston, October 29, 1973, the Citizens National Bank records in Austin show that Jacobsen entered box 865 at 2:00 P.M. Likewise, bank records show that on Monday, November 26, 1973, the day after he met Connally at George Christian's house, Jacobsen's former law partner, Joe Long, had access to safe deposit boxes at the Citizens National Bank, and Long confirmed that Jacobsen had access to his two boxes at the same time. The prosecutors found persuasive the fact that Jacobsen had access to the boxes the day after meeting Connally, particularly because Jacobsen was not aware that the FBI was going to inventory the money the next day:

> Obviously, if Jacobsen, on his own, thought he should change the money, he could have done so at any time between November 2 and November 26, a period of fifteen business days. He chose, however, to enter on November 26, just the day after the undisputed evidence shows that he had seen Connally alone. The inference arising from this pattern of seeing Connally alone and then going to the box is plain. Connally gave something to Jacobsen.[82]

THE OCTOBER 29 MEETING

Another persuasive piece of circumstantial evidence to the prosecutors was Connally's meeting on October 29 with Jacobsen, when he paid for Jacobsen to fly from Austin to Houston ostensibly to discuss with him the bank application of Connally's client, Gus Wortham. The prosecutors thought there were a number of holes in this story. For one thing, even if Wortham was an important client and Connally wanted the matter taken care of before he left for a two-week trip to Europe, it did not explain "why Connally had to see Jacobsen rather than talk to him over the phone, especially since Connally said he did not give Jacobsen any documents and wanted to save time." Further, the prosecutors reasoned that if Connally really wanted the Wortham matter taken care of before he left for Europe, it was likely that he would call Jacobsen upon his return from Europe to find out where the matter stood. Connally did *not* do so. Finally, Connally's excuse that he wanted to see Jacobsen personally on October 29 because he had heard and observed that Jacobsen was being adversely affected by his personal financial problems—"in short, Connally wanted to judge [Jacobsen's] demeanor and not just hear his voice"—was unconvincing. For one thing, Connally had testified earlier that "the" reason that he wanted to see Jacobsen was to save time. Further, as the prosecutors observed:

> It is at total variance with Connally's own acknowledgment that he saw Jacobsen just three days earlier on October 26 and in fact was sufficiently perceptive of Jacobsen so that he sensed that when Jacobsen offered to lie, he was just feeling Connally out. In short, this "reason" which Connally testified to is a demonstrably false exculpatory statement.[83]

In the last analysis, therefore, what led the prosecutors to proceed with the indictment was less Jake Jacobsen's testimony alone than it was John Connally's own weak, contradictory and inconsistent explanations, coupled with the circumstantial evidence supporting Jacobsen. The prosecutors never did satisfactorily answer for themselves the mystery of the third $5,000. Nevertheless, they felt the mystery was outweighed by Connally's own weak story.

The third $5,000 was given to Jacobsen in October of 1971. Connally's testimony was that Jacobsen had twice offered him $10,000, once in June of 1971, and again in the summer of 1972. The rationale of the prosecution was that if you believed the anticipated defense that Jacobsen had pocketed all of the $15,000 himself, why would Jacobsen (according to Connally's testimony) have bothered to go back to Connally in the summer of 1972 and again offer $10,000 to Connally. And if Jacobsen really had $15,000, why did he offer just $10,000 in the summer of 1972. Indeed, if he had pocketed the entire $15,000, why did he make *any* offer in the summer of 1972? Since no new money had come to Jacobsen from the dairy lobby since November of 1971, why did he wait until the summer of 1972 to make another offer? If the dairy lobby had given the third $5,000 to Jacobsen in the summer of 1972, it might be a different story; but the dairy lobby had not.

So said the prosecution. But they were playing a shell game of sorts. Ignore the weaknesses of Jacobsen's testimony on the third $5,000. Look instead at the weakness in Connally's testimony. But the weakness in Connally's testimony did not automatically strengthen Jacobsen's testimony.

After Jacobsen testified at the trial, the prosecution called a parade of witnesses from the twelve Federal Reserve Banks to prove that all of the "October 29 money" had actually been in circulation on that date, despite the fact that the government did not have possession of it until February 24 of the next year. Edward Bennett Williams' low key cross-examination attempted to show that some of the money might not have been in circulation on October 29. These witnesses were crucial to the prosecutors because this was the essential element which convinced them of Jacobsen's veracity. Only one bill, one bill which had not been in circulation on October 29, 1973, would have made him a liar, but such a bill was never found. All of the bills corroborated Jacobsen's story. By most accounts, however, observers were not nearly as impressed as the prosecutors. Said Judge Hart, "I don't understand all this, bless me, I don't."[84]

In all, Tuerkheimer called thirty-five witnesses before he stated to the court, "The government rests." As expected, Williams filed a motion to dismiss the charges against Connally and, as expected, Judge Hart denied the motion. Williams thereafter led off with a host of character witnesses, Democrats all: Lady Bird Johnson, former Defense Secretary Robert McNamara, and former Secretary of State Dean Rusk, Robert Strauss, chairman of the Democratic National Committee, and Congresswoman Barbara Jordan from Texas. They all thought Connally was a swell guy with honesty and integrity as his constant qualities.

The best, however, was yet to come: Richard Nixon's favorite preacher, Dr. Billy Graham, who testified that he was "an evangelist preaching the gospel of Jesus Christ over the world." "Amen" whispered several of the jurors to themselves. Dr. Graham testified that he had known John Connally for twenty-one years and had first met him when "I was in Texas reading the Bible to a friend (oil baron Sid Richardson), and he said, 'Wait a minute. I'd like to have my lawyer hear this,' and in walked John Connally." And Dr. Graham had known Connally ever since. "He's spoken from the platform in two of my crusades, and has attended others. In Washington and Texas we sometimes played golf . . . and we always have a little prayer together."[85]

Connally himself was Williams' last witness. Blue pinstripes. Silver hair. The commanding presence. Just the right touch of humility at finding himself in such a position. Williams took him through his direct testimony. His story was repeated yet another time. Tuerkheimer's cross-examination essentially did no damage. At one point, however, Tuerkheimer asked Connally a curious question. Had Connally ever offered Nixon "a very substantial allocation of oil in Texas." "No," said Connally, they were discussing revenue sharing, and he told Nixon "a substantial allocation of taxes" would be at his discretion.[86]

There was never any follow-up on this question by Tuerkheimer, and journalists at the trial reported that the jury did not appear to understand the exchange. Judge Hart, however, certainly understood it and had deliberately limited Tuerkheimer's questions on the subject. The story involved a separate meeting which Nixon and Connally had had alone on September 23, 1971, immediately after Connally's meeting with the president and other advisers on the milk price support decision. The special prosecutor's office made a transcript of that conversation from a tape which was barely intelligible. That transcript was never introduced into evidence, and Williams persuaded Judge Hart to suppress it and seal it from public view. A copy of this transcript has since been made public:

Connally: "May I have two minutes with you on another matter?"

President: "Sure, sure, sure. Sit down. Absolutely. . . ."

Connally: "I think you made the right decision."

President: "They are tough political operators. This is a cold political deal."

Connally: "It's on my honor to make sure that there's a very substantial allocation of oil in Texas that will be at your discretion."

President: "Fine. . . ."

Connally: "Unless you want somebody else to do it. Somebody. . ."

President: "No. . . ."

Connally: "I'm sorry to bug you about it, Mr. President."[87]

The portion of the tape from which the prosecutor's transcript is quoted above was, at Williams' suggestion, prepared but not played for the jury until the end of the trial. Given the unintelligible nature of the tape, the absence of a transcript, and the lack of awareness of its significance, it breezed right past the jury. And Connally's dubious explanation was never challenged. Taxes at Nixon's discretion? In connection with revenue sharing? Since when does revenue sharing put taxes at a president's disposal? What's even more curious is that "oil" does not even *sound* like "taxes." The special prosecutor's office believed that Connally's reference to "oil" really meant "cash," i.e., that AMPI had more money waiting in Texas for Nixon to use. They also believed the reference tied Connally in more deeply to the price support decision. Said Tuerkheimer in the judge's chambers:

> I think the tape shows that he followed it through, Secretary Connally did, from the very beginning, that he pushed for it and argued for it, and he saw it through to a final decision by the president. . . . Here is what happened: when Mr. Connally first appeared in the grand jury he said he didn't learn about the decision until he read about it in the papers. Then when we played the tape recording for him, that is, as much of the tape recording as we had, he still took the position, well, the decision wasn't final, it might have gone another way. . . . He tried, as far as I could see, to make the decision something less than final and I think that the last segment here makes clear beyond any doubt that as far as he knew this was a final decision, he cemented it with the president and that, I submit, is inconsistent with the position he has taken in the past, and it deals with the element of the crime. It shows the issue was raised at Connally's instance. He says, "I am sorry to bug you about it, Mr. President."[88]

Judge Hart was ambivalent:

That part of it which I can easily understand from that tape is what starts out, "May I have two minutes with you on another matter?" And at least to me, that is clearly Connally's voice. . . . Then it was clear Connally said, "It's on my honor to make sure there is very substantial oil in Texas that would be at your discretion." . . . Then this oil-in-Texas business could be interpreted as something else, or referring to campaign contributions. I think. And then over again, it was quite clear that he said something here. "I think you made the right decision." Again I think the jury could come to the conclusion that he was talking about milk support prices.[89]

Williams did not want the tape admitted under any circumstances:

I have listened to it now several times . . . and I say to you, Your Honor, that it is incomprehensible, that its meaning is incomprehensible. . . .[90]

And Judge Hart eventually agreed with Williams:

Well, I think under the circumstances I will rule it out in direct examination, on your case in chief.[91]

For good reasons, juries are not permitted to know many things about a defendant because of the prejudicial effect such knowledge might have on their deliberations. While there were weaknesses in Connally's testimony, the jury did not know these "weaknesses" had resulted in Connally's indictment on two counts of perjury and two counts of giving false declarations to a grand jury. Also, while the jury knew that Connally and Jacobsen had both failed to tell the grand jury about their October 26 meeting at the Sheraton and that Connally had sent copies of the transcripts of his appearances before the Senate Watergate Committee and the Watergate Special Prosecution Force, as well as a detailed digest of his November grand jury appearance, to Jacobsen, the jury did not know that Connally had been indicted for conspiring with Jacobsen to obstruct justice and make false declarations before the grand jury and Senate Watergate Committee.

These are important distinctions to keep in mind when analyzing both the jury's not-guilty verdict on the two counts of receiving an illegal gratuity and the foreman's cryptic comment that their verdict did not necessarily mean that they had found Connally innocent, "but rather not guilty on the case presented to us." Connally never stood trial on the lesser counts where he was possibly in greater jeopardy.

Consider Jacobsen's story once again. The prosecution claimed in its closing argument to the jury that "on virtually every conceivable point where Mr. Jacobsen could be corroborated, he has in fact been corroborated." To an extent, this was true. Jacobsen's story of delivering the money to Connally in 1971 and Connally's delivering the money back to him twice in 1973 is corroborated both by the safety deposit box pattern and Connally's weak explanations of the purpose for the last 1971 meeting and the two 1973 meetings. Jacobsen's story of originally getting the money from the dairy lobby for delivery to Connally is corroborated by Bob Lilly's testimony.

But even if the jury believed Jacobsen's story on those points where it was corroborated, they still did not have to find John Connally guilty of receiving an illegal gratuity. Why? Because the one meeting between Connally and Jacobsen which did not have, and by its nature could not have, any corroborating evidence was the one on April 28, 1971, when Jacobsen claimed that Connally had asked him for some of the dairy lobby's "political money" because he had been "of help on that milk producers thing," allegedly a reference to his work on the 1971 milk price support decision.

There obviously could be no corroborating evidence on the words exchanged between the two. Jacobsen was alone against John Connally's credibility. The prosecution hoped that all that followed afterwards, Jacobsen's call to Lilly later on April 28 asking for money for Connally, the suspicious meetings, the safety deposit box patterns, the issuance dates of the money, and Connally's weak explanations and contradictions, would lead the jury to accept Jacobsen's version of what happened at the April 28 meeting.

It didn't work. The jury did not believe Jacobsen; they did not believe that John Connally could bring himself to ask for money and crudely remind Jacobsen at the same time of what Jacobsen already knew, i.e., that Connally had helped the dairy lobby enormously on Nixon's 1971 price support decision. It simply did not ring true. What happened after the April 28 meeting was in a sense irrelevant to the illegal gratuity charges. Everything after that could have happened just as Jacobsen claimed, and Connally still would not have been guilty on those charges.

We can never know what the jury really believed happened on April 28, 1971. Even if individual jurors came forward and attempted to answer that question now, they could speak only for themselves. They could not know what their fellow jurors were thinking at the time, and indeed they may have forgotten by now what they themselves were really thinking at the time.

What the jury *could* have believed about the April 28 meeting is that "political money" *was* discussed, if only because Jacobsen called Lilly the same day and asked him for money to give to Connally. What they also could have believed was that, as Connally testified, Jacobsen offered Connally the money solely for his use as political contributions without any crude, explicit references to the recent price support matter. And what they could have believed was that, contrary to Connally's testimony, Connally agreed to take the money for its intended use—political contributions.

The jury could have believed all of these things. And if they did, then John Connally would still not be guilty of receiving illegal gratuities, and all the other evidence against him would not have mattered because the jurors did not have before them the conspiracy, perjury, and false declaration counts.

In a very real sense, the most serious charges against John Connally—receiving illegal gratuities—were the weakest because in order to find him not guilty, the jury did not have to believe that Connally never received the money. All they had to believe was that two shrewd political operators like Connally and Jacobsen would not be so naive as to discuss official acts in the same breath as receiving political cash from the beneficiary of those acts.

While there was evidence tending to show that Connally and Jacobsen had engaged in a conspiracy to obstruct justice by their testimony before the grand jury and the Senate Watergate Committee, the jury never had to evaluate John Connally's culpability on such a conspiracy count. And today, John Connally is a free man.

Whatever happened to Jake Jacobsen? Columnist William Safire made a prediction before the trial:

> Whichever way it goes, it's jake with Jake. Thanks to the special prosecutor, Mr. Jacobsen is out of his big Texas fraud trouble and has pleaded guilty to only giving a bribe. If the jury

does not believe that he bribed Mr. Connally, . . . then no bribe
was given and Mr. Jacobsen has his freedom and a good long
laugh.[92]

Bill Safire, however, was one of Nixon's speech writers and not one
of his lawyers. Connally may have been innocent of accepting a
bribe from Jacobsen, but Jacobsen had pleaded guilty to bribing
Connally and he stayed guilty. Connally's acquittal did not take
Jacobsen off the hook. Over a year later on August 21, 1976,
Jacobsen appeared, bankrupt, before Judge Hart. Charles McNellis,
Jacobsen's lawyer, had requested probation for his client because
the severe illness of his wife would require placing her in a public
institution if Jacobsen went to prison. Jacobsen threw himself "on
the mercy of this honorable court" and asked that the court "take
into consideration the physical and mental condition of my wife
and my financial condition." Hart did:

> You have committed crimes that severely undermine the very
> foundation of this Republic and its judicial system. You have
> walked with princes in your time and you have been reduced to a
> state of degradation and disrepute that is the lot of a common
> criminal.
>   . . . Whatever the many errors of your ways, you have been
> more than a model of conjugal devotion.[93]

Hart agreed with the pleas of Jacobsen's lawyer since he had read
the doctors' reports submitted to him indicating that Jacobsen's
wife had been severely ill for many years and that being placed in a
public institution should Jacobsen go to prison "may well be fatal"
to Mrs. Jacobsen. Hart then sentenced Jacobsen to the maximum
prison term of two years and $10,000 fine and then suspended the
sentence.[94] Nobody laughed.

The only humor was furnished by the dairy lobby. No matter how
bitter the disputes had been between prosecutors and defense
counsel, no one denied that the dairy lobby had willingly and
knowingly *attempted* through Jake Jacobsen to bribe the secretary
of the treasury. Unashamed, AMPI even asked the government to
return to it the $10,000 which was used as evidence at the trial. Out
of the entire story of the attempted bribery of John Connally, only
that deserves a good, long laugh.

8

Son of ITT:
The AMPI
Antitrust Suit

For most Americans, Richard Nixon's resignation signalled the end of Watergate. The dairy lobby in general and its largest member AMPI in particular were not so fortunate. As we have seen, their attempt to bribe John Connally kept them publicly involved in Watergate-related matters through the spring of 1975. Shortly thereafter, U.S. District Court Judge John W. Oliver, the chief judge for the Western District of Missouri, approved and signed a consent decree in the federal government's three-year-long antitrust case against AMPI, thereby terminating the dairy lobby's last contact with Watergate.

The AMPI case was the first of three government antitrust cases brought against the dairy lobby's major members in the early seventies. Cases were also brought against Mid-Am and DI. From its inception to its conclusion, however, AMPI's case was the one intimately tied to the Watergate atmosphere.

Throughout the late 1960s and early 1970s an increasing number of complaints about AMPI and its activities were brought to the attention of the Antitrust Division of the U.S. Department of Justice. In February of 1971, John Sarbaugh, chief of the Chicago office of the Antitrust Division, was authorized to conduct a preliminary inquiry into AMPI's price-fixing of raw milk in the Dallas, Texas, area.[1]

By August 1971, the Antitrust Division had completed its preliminary investigation. The Chicago field office believed the AMPI activities it had uncovered were so blatant and heavy handed in terms of price fixing, monopolization of many regional markets, and "conspicuously predatory behavior" that a full scale criminal grand jury investigation was warranted.

The grand jury recommendation was approved at several levels in Washington until it reached the desk of Richard McLaren, the assistant attorney general in charge of the Antitrust Division and a Nixon political appointee. McLaren, a respected antitrust lawyer and law professor, also supported the request for a criminal investigation. He agreed with the assessment made by the chief of the Justice Department's General Litigation Section:

> if the grand jury investigation were to convince us that AMPI is not acting as badly as we think it is or that it was undertaking these practices with the bona fide belief that they were not illegal, we could then decide the criminal prosecution would not be appropriate, but on the facts available to us now, I think that a grand jury would be justified.[2]

As a political appointee, McLaren was aware that Nixon had recently spoken to AMPI's second annual convention in Chicago in front of forty thousand members and their wives. To cover himself, McLaren talked to the U.S. Department of Agriculture to see if they had any objection to a criminal investigation of AMPI. The Department of Agriculture had no objections and McLaren's contact there "pretty well confirmed that AMPI has some very rough characters and there is fire under the smoke."[3]

Accordingly, McLaren sent a routine memorandum to Attorney General John Mitchell on September 9, 1971, recommending a grand jury investigation. McLaren normally received a response from Mitchell on such matters within a week or two. When Mitchell had still not indicated his preference by the end of October, McLaren sent Mitchell another note requesting authority for a grand jury and stressing the continuing nature of AMPI's alleged illegal tactics.[4]

During the same period, AMPI was hardly unaware of the Justice Department investigation, and neither was the White House. After McLaren had passed on his first grand jury recommendation to

Mitchell, Charles Colson sent an "eyes only" memorandum to Haldeman on the subject:

> For obvious reasons, I should not be involved with respect to the following. There is underway in the Justice Department at the moment an Antitrust Division investigation of the milk producer cooperatives. Attached is the 1956 court decision exempting the milk producers from application of the antitrust laws. *If this goes too far there may be a number of very serious adverse consequences which I would be glad to elaborate on in detail.*[5]

Haldeman reacted positively to Colson's memo and had his aide Gordon Strachan contact White House counsel John Dean. Strachan asked Dean to check into the Antitrust Division investigation "on a very low-key basis."[6] Dean subsequently prepared a report for Strachan on "the current activities within the Department of Justice regarding the Association of Milk Producers, Inc., [sic]" and delivered it early in October.[7] Haldeman was disturbed by Dean's report that such a large contributor to Nixon's re-election was facing criminal antitrust charges. There was already enough adverse publicity tying Nixon to "milk money."

Throughout the spring of 1971, in a series of exclusive stories for the *Minneapolis Tribune,* Frank Wright, its Washington correspondent, had been systematically exposing the political contribution activities of the dairy lobby. While the Nixon White House had been worried by Wright's investigative reporting, his efforts were not receiving much national publicity. Despite the Nixon efforts to obfuscate the dairy contributions through a series of sham committees and a multiplicity of checks and treasurers, the dam finally broke in late September 1971 with a *Wall Street Journal* story headlined "Milk and Money, Flood of Cash to Help Nixon Campaign Follows Hike in Dairy Support" and a story in the *Washington Post* headlined "Milk and Dollars for Nixon—Dairy Cash Pours into Dummy Committees," both belatedly covering the same ground plowed earlier that year by Frank Wright.

Given the adverse publicity Nixon was beginning to receive, Haldeman decided to meet with Mitchell to see what could be done to help AMPI. The "talking paper" for the Mitchell meeting prepared for Haldeman by Gordan Strachan thoroughly covered the interwoven dilemmas of the AMPI antitrust investigation and noted that the *Washington Post* had assigned four reporters full-time on the milk money project.

Mitchell and Haldeman met on November 4, 1971, approximately a week after McLaren's second request to Mitchell for a criminal grand jury. No record exists of precisely what was discussed between them but Strachan's talking paper makes clear that the antitrust investigation of AMPI was considered in the context of Nixon's campaign and the contributions to it from the dairy lobby.

What is also clear is that Haldeman either then or later secured Mitchell's agreement to drop the criminal investigation of AMPI. Mitchell met with McLaren at the end of November and gave him the word: despite the recommendations of everyone in the Antitrust Division, from the attorneys in charge of the Midwest Field Office right up to the head of the Antitrust Division, there was to be no criminal grand jury investigation. The most Mitchell, who had no experience as an antitrust lawyer, would approve was a civil action.

McLaren knew he was up against the realities of AMPI's political influence with Nixon and moved quickly to avoid having the approved civil action outflanked as well. On December 20, 1971, the midwest office of the Antitrust Division prepared a draft complaint for filing against AMPI, and the complaint, which was reviewed through the normal chain of command, was unanimously approved. On January 18, 1972 McLaren also showed the proposed complaint to Richard Laing at the U.S. Department of Agriculture, who said he had no objection. Following that meeting, the complaint was sent to Mitchell with a strong recommendation urging approval. Within a few days, on Saturday, January 22, Mitchell approved filing the complaint. Once again, however, Mitchell played politics. He requested that the Justice Department give AMPI the opportunity to enter into negotiations for a consent decree before the lawsuit was filed. Under this procedure, the defendant is shown the civil complaint *before* it is filed in court. If the defendant essentially agrees not to do again what the civil complaint charges him with, the government and defendant enter into a "consent decree," which is filed simultaneously with the complaint.

During this same period when the complaint was scrutinized by him for approval, Mitchell was actively involved in directing political contribution/solicitation activities from AMPI as well. On January 13, 1972, Mitchell met Kalmbach to discuss Kalmbach's efforts to have AMPI fulfill its $2 million commitment. Kalmbach reported back to Mitchell on his progress with that commitment the day before Mitchell told McLaren to delay filing the lawsuit and give AMPI a chance to negotiate.[8]

During the Senate Watergate Committee investigation, McLaren, in a letter dated May 10, 1974 to David Dorsen, assistant chief counsel for the committee, stated that he did not feel it was "unusual" for Attorney General Mitchell not to approve recommendations from the Antitrust Division although McLaren did "not recall any [other] specific cases in which this occurred."[9]

McLaren, however, was less than candid. Mitchell was up to his ears in milk money and he was not about to risk sending such a major contributor to jail. A civil action, if Justice used the traditional and cumbersome Civil Investigation Demand (CID) to obtain its evidence before filing, could tie up the matter for a year or more. Mitchell knew this and probably was counting on it. McLaren knew it as well, which is why he moved quickly to forestall any further political interference from Mitchell. Rather than go the CID route, he essentially told his staff to conduct a few more witness interviews and then go straight to court. Again, however, Mitchell moved to thwart McLaren by asking for prefiling negotiations—clearly another stalling tactic. AMPI certainly saw this as an opportunity to delay, and hence the course of the prefiling negotiations was hardly placid.

On Monday, January 24, McClaren told John Sarbaugh to notify AMPI of the complaint and "offer AMPI the opportunity to engage in prefiling negotiations." Sarbaugh phoned AMPI attorney Stuart Russell that day and gave him until the close of business on Thursday, January 27, to decide whether AMPI wished to engage in prefiling negotiations. The policy was, Sarbaugh informed Russell, that

. . . prefiling negotiations are offered by the department in situations where the defendant has agreed in principle to the relief requested by the department; that under division practice after agreeing to the prefiling procedure, parties have sixty days to negotiate a definitive consent decree; and that prefiling is not engaged in when there are genuine issues of law or fact in dispute between the department and the defendant."[10]

In other words, if you agree that we have you cold on the facts and the law, we will try to work out a consent decree. If you protest your innocence or argue about our view of the law we will see you in court.

Russell shortly thereafter phoned AMPI attorney, Marion Harrison, and told him of the conversation with the Chicago office of the Antitrust Division and the lack of advance notice for the action, complaining that this was ". . . contrary to the usual modus operandi involving antitrust suits." Marion Harrison agreed. The action

> . . . had just come out of the blue. . . . The Justice Department, contrary to its usual practice, was allowing AMPI a brief period of time, which I think was forty-eight hours—at any rate it was a very short period of time—to consent to a proposed consent decree or the lawsuit would actually be filed, and was not allowing AMPI its own copy of the proposed complaint in the lawsuit, but was limiting it to coming to the Chicago office and reading the complaint there.[11]

It *was* unusual. But then most potential antitrust defendants do not promise to contribute $2 million to the president's re-election. McLaren had simply once again moved to check Mitchell's political interference by providing only a strictly limited opportunity to AMPI for prefiling negotiations.

On January 25, AMPI Chicago counsel Erwin Heininger came to the midwest office of the Justice Department to read the complaint and, on the next day, advised the government lawyers that his client did wish to engage in prefiling negotiations. However, in his report to McLaren, Sarbaugh told him that AMPI lawyers had also questioned the propriety of filing the complaint at all—emphasizing the antitrust exemption of the Capper-Volstead Act, and AMPI's new management and new attorneys. The AMPI lawyers further made a point of noting that McLaren would shortly be leaving the Antitrust Division to become a federal judge and that Attorney General Mitchell might be on his way out also, thus "necessitating new signatures if the complaint against AMPI were to be filed at a later date."[12] Finally, Sarbaugh also reported that, in a particularly heavy handed manner, AMPI lawyers had mentioned that their clients "were big political contributors." Sarbaugh, who had known Erwin Heininger, AMPI's Chicago antitrust lawyer, on a professional basis for several years ". . . was shocked and concerned by the reference to political contributions." He thought that this was the most "blunt and tactless" comment he had ever heard Heininger make.

Disheartened by AMPI's attitude, Sarbaugh told its counsel again on January 27 that prefiling negotiations were entered into only within certain limits and that if there were "genuine issues of dispute" necessary to be resolved or contested, prefiling negotiations were not appropriate, and a breakdown in the prefiling negotiations would result in their being terminated. While AMPI lawyers again stated their willingness to enter into the negotiations and represented that there was "sufficient agreement to warrant prefiling negotiations," Sarbaugh was suspicious that "AMPI might . . . use the prefiling negotiation period to attempt to block politically the filing of this suit."[13]

Sarbaugh's report greatly concerned McLaren. He told Sarbaugh to prepare a draft consent decree and submit it to AMPI counsel at the close of business on Friday, January 28. Pursuant to McLaren's instructions, AMPI was told that it had only until the close of business the next Monday to inform the Justice Department if AMPI would "consent in principle to the basic prohibitions in the proposed decree"—barring which the complaint would be filed by February 1—McLaren's last day as assistant attorney general.[14]

AMPI was dismayed by this short notice, especially since AMPI's president did not receive a copy of the proposed decree until the Monday morning deadline for its acceptance. When AMPI could not agree to the decree on that day, McLaren ordered that suit be filed the next day, Tuesday morning, February 1, 1972, and it was.[15]

In a letter to the Senate Watergate Committee staff, McLaren conceded that he ordered the complaint filed on February 1 ". . . partly to wrap up unfinished business and partly to preclude any possible attempt by AMPI to resist the filing of the complaint by some political means, since representatives of AMPI had indicated to Mr. Sarbaugh their consideration of political factors." McLaren's successor as assistant attorney in the Antitrust Division, Thomas Kauper, subsequently stated that "it must be conceded that this was somewhat unusual. McLaren, I suspect, feared there might be pressure *not* to file."[16]

The first month of 1972 was a strain for AMPI. The antitrust litigation was dropped in its lap less than two short weeks after a major change in its internal organization, and about the same time consumer crusader Ralph Nader decided to go to court over Nixon's milk price support reversal in 1971.

AMPI was reportedly near bankruptcy in early 1972, when its board of directors replaced Harold Nelson as general manager with Dr. George Mehren, an agricultural economist, former Department of Agriculture bureaucrat and AMPI consultant. The dairy lobby takes care of its own, however, because Nelson was merely kicked upstairs to be a consultant to AMPI at $100,000 a year. Dave Parr, who also lost his job with AMPI at the same time, soon landed on his feet with a position at DI, the dairy lobby's third largest member.

During his first weeks on the job, Mehren, a lifelong Democrat who had been assistant secretary of agriculture in the Johnson Administration from 1963 to 1968, made an effort to determine the various commitments the prior management had made. One of his major priorities was cutting expenses, and he was most interested in what political cash promises had been made by his predecessors.[17] In John Butterbrodt's presence, Mehren asked Dave Parr about such commitments. At the time, "only" $1.3 million existed in funds for political donation, and Parr answered that there were no existing commitments, neatly avoiding any reference to the dairy lobby's overall commitment of $2 million to Nixon, $1.75 million of which was still unfulfilled. The extent of Parr's candor on this occasion can be judged by his further claim that he did not know a Mr. Colson at the White House.[18]

Indeed, the same day Mehren was taking the helm of AMPI, Harold Nelson and Jake Jacobsen were on their way to California for a previously scheduled meeting with Nixon's chief fund-raiser, Herb Kalmbach. Kalmbach bluntly announced that there were commitments from the dairy lobby to Nixon which must be carried out. Nelson, for his part, told Kalmbach both about the AMPI organizational changes taking place that day and AMPI's continued desire to make a substantial commitment to the Nixon effort. Nelson refused to commit to a specific amount since he was no longer in control of AMPI, a fact not even brought to Jacobsen's attention until they both boarded the plane to California.[19]

George Mehren, however, quickly learned of the dairy lobby's commitment to Nixon, despite Parr's denials to the contrary. As a consequence, a second meeting was soon scheduled for Kalmbach to meet again in California with Jacobsen, Nelson, and Mehren.

The meeting took place on February 2, 1972, at a club in Newport, California, the day after McLaren had filed the government's antitrust action against AMPI. If this affected Kalmbach at all, he did not show it. He was interested in pinning Mehren down on a precise figure and a timetable for payment, something he had not been able to do with Nelson at their previous meeting. Kalmbach promptly asked Mehren "when the hell" the money was coming and stated that the money was "owed"—something which Mehren vehemently denied, leaving Kalmbach unhappy.[20]

Nevertheless, during the course of the meeting, a number of things happened. Mehren specifically raised with Kalmbach the question of the antitrust suit and complained about the poor treatment AMPI was receiving from the administration. According to Jake Jacobsen, Mehren even suggested that further contributions by the dairy lobby might "alleviate their problems with . . . the antitrust suit."[21] Kalmbach was sympathetic but did not agree to talk to anyone in the administration about the suit.

Nevertheless, dairy lobby commitments were tentatively reconfirmed, although the $2 million pledge was reduced to $1 million. After the renewed pledge, the discussion then centered around which part of the contribution would be made before April 7, 1972, the effective date of the new federal campaign reporting statute. While Jacobsen had trouble remembering the exact figures, he conceded that Kalmbach's proposition might have been that any contribution should be made "one-third, one-third, and one-fourth before April 7 and the remainder after April 7. . . ." This sounded fine to Jacobsen, but Mehren did not firmly commit himself to the schedule or the $1 million amount.[22]

Kalmbach was clearly uncomfortable with Mehren's mention of the antitrust problems of the dairy lobby in the same meeting at which campaign contribution pledges were renewed and time tables for payment discussed. He was also concerned about Ralph Nader's lawsuit and the risk of being subpoenaed to testify in that action. Promptly after his meeting with Mehren, Nelson, and Jacobsen, therefore, he conveyed his fears to Gordon Strachan, who then recommended to his boss, H.R. Haldeman, that Kalmbach be relieved of dealing with the dairy lobby.

After their early February meeting, Mehren had Jacobsen continue to keep in phone contact with Kalmbach. The dairy lobby officials were unaware, however, that Kalmbach was doing his best to end his fund-raising relationship with them. In this regard, Kalmbach received an unexpected break in the form of the ITT-Dita Beard antitrust scandal, which broke into public view via a Jack Anderson column late in February.

Since the ITT scandal also involved alleged manipulation of an antitrust action in exchange for campaign contributions and other considerations, the White House suddenly became far more responsive to Kalmbach's desire to be rid of the dairy lobby, its money, and its incessant whining about the unfairness of the AMPI antitrust suit. Two antitrust campaign contribution scandals at once were more than Nixon wanted to risk in those pre-election days. Kalmbach was thereafter authorized through Haldeman to terminate his dealings with the dairy lobby. A meeting was arranged in Washington at the Madison Hotel on March 16, 1972, with Jacobsen, Nelson, and Mehren. Unaware of the change in Nixon's attitude towards them, the dairy lobby officials met there and told Kalmbach that AMPI had finally decided to make substantial contributions to the Nixon campaign. Kalmbach politely listened to them and then dropped his bombshell: he no longer wished to receive contributions from AMPI. The dairy lobby's reaction was severe—"You talk about shock," said Nelson. Jacobsen was also surprised and remembers that Kalmbach "was in such a hurry he just gave us the decision and kind of grabbed his bags and went out."[23]

Kalmbach explained later what his feelings were at the time:

> I felt the publicity for the campaign, the negatives of that far out-weighed the actual funds received and I went into that meeting with two purposes in mind, and I made up my mind I was only going to be there for five or ten minutes. One, that I was going to tell them that, as far as I was concerned, we were not interested in receiving any more funds from AMPI . . . and second, if they felt they had a pledge outstanding to the campaign, that that pledge was, in fact, abrogated. And I think I did that.[24]

The crestfallen trio (Mehren, Nelson, and Jacobsen) had another significant meeting that same afternoon, this time with Secretary of the Treasury John Connally. Connally gave them no more real comfort than Kalmbach, although he discussed with them the topics of campaign contributions, the antitrust suit, and a recent tax problem which had arisen from AMPI's past.

According to Connally, the main purpose of the March 16 meeting was to inform him of the AMPI reorganization, which left Mehren in charge. Although the AMPI representatives told Connally about a number of their problems, including those relating to marketing practices and the antitrust suit, Connally has denied that he was asked to do anything specific.

At one point during the meeting, after the question of campaign money was raised, Connally turned to Jacobsen and said, "Jake, you've got lots of problems at AMPI [and] we would just prefer not to take any money from you." Mehren, apparently unable to comprehend that the Nixon Administration now considered the dairy lobby a liability, suggested that perhaps under the circumstances, AMPI should delay making any further contributions until later in the campaign when "the heat [was] off."[25] Connally agreed that this was reasonable.

Mehren was still unwilling to accept as fact the brush-off he received from both Kalmbach and Connally. Three weeks later, he was arranging for the dairy lobby to make substantial contributions once again to the Nixon campaign in partial fulfillment of the tentative timetable agreed to in his earlier February meeting with Kalmbach. Mehren did not know it at the time but his actions on that occasion almost got him indicted by the Watergate grand jury.

On April 4, 1972, Mehren called AMPI's president, John Butterbrodt, who was attending another dairy cooperative's convention
in Wisconsin, to receive his approval for issuing thirty separate
$5,000 checks from TAPE for subsequent contributions to the Re-
publican party and Nixon's re-election campaign. Next, Mehren
called AMPI attorney Jake Jacobsen in Austin, Texas and told him
that he wanted to talk to Herb Kalmbach right away. Jacobsen
promptly contacted Kalmbach who agreed to call Mehren at his
home in San Antonio that evening. Jacobsen called Mehren back
and told him when to expect the call.[26]

Harold Nelson asked Mehren what he expected to gain by talking with Kalmbach. Mehren responded as if he had not been
listening to Kalmbach or Connally three weeks earlier. He told
Nelson that he wanted Kalmbach and the Republicans to know that
AMPI was not welching on its commitment to the Nixon re-election
and that AMPI also expected the Justice Department to slow down
its antitrust suit against AMPI and later reduce it to a wrist slap.[27]

In the meantime, after receiving Butterbrodt's approval for the
TAPE campaign contributions, Mehren discovered that there were
only twenty-six checks left in the TAPE checkbook in San Antonio,
not enough for the thirty separate $5,000 checks which Butterbrodt
had approved. Jacobsen was again called in Austin and asked to
have Citizens National Bank (where Jacobsen was Chairman of
the Board) deliver four additional bank checks from Austin to San
Antonio *that same day*. Later that afternoon, John Parker of Citizens National Bank called Bob Lilly at AMPI's San Antonio office
to advise him that a bank employee would be personally delivering
the additional four checks to AMPI's office around 6:00 P.M.[28]

When the extra four checks arrived from Austin, all thirty of
them were then made out for $5,000 each and dated April 4, 1972,
with the names of the payees left blank. All the checks were signed
by AMPI comptroller Lynn Elrod, who had them delivered to
Mehren's home that same evening for his signature. Mehren had
told Lilly earlier that day that the names of the payees would be
furnished later by either Kalmbach or John Connally. Mehren also
indicated to Lilly that the Mid-America Dairymen's political action
committee, ADEPT, was going to deliver $100,000 in checks to

Nelson, and Dairymen, Inc.'s political action committee, SPACE, would contribute $50,000.[29]

According to Kalmbach, he did call Mehren on the night of April 4. Mehren was frustrated and let Kalmbach know it. Mehren told Kalmbach of his concern about the antitrust situation and his desire to have Kalmbach intercede on AMPI's behalf with the White House on the antitrust problem. At the same time, Mehren let Kalmbach know that AMPI was getting ready to make another large contribution to the Nixon campaign immediately prior to April 7, 1972, the date when the new campaign disclosure statute became effective. Kalmbach, however, gave Mehren no more solace this time than he had during their brief meeting on March 16. He bluntly told Mehren that he would not intercede with the White House and would not do anything to assist AMPI in its antitrust difficulties. Upon being turned down by Kalmbach, Mehren abruptly terminated the telephone conversation and subsequently voided all thirty of the $5,000 checks which he had gone to so much trouble to have issued and signed on April 4. Thus, the thirty checks were never delivered to Nixon or the Republicans.[30]

On the basis of Kalmbach's version of these actions of Mehren on April 4, 1972, and Mehren's later testimony to the grand jury regarding his conversation with Kalmbach, the Watergate Special Prosecution Force believed that Mehren had committed perjury when he denied, under oath, seeking Kalmbach's assistance in ending the antitrust suit in exchange for campaign contributions.[31]

In Mehren's favor, however, was the fact that Kalmbach, who the prosecutors believed was being scrupulously honest in his dealings with them, could not testify that Mehren *explicitly* tied his offer of making substantial Nixon campaign contributions to his request for Kalmbach to assist him in intervening with the White House on the antitrust suit. Kalmbach said that the offer of the contribution did immediately follow Mehren's request for intercession with the White House, and Kalmbach assumed that the two matters were joined together.[32]

Nevertheless, the prosecutors were further convinced by circumstantial evidence—the trouble Mehren took on April 4 to issue and sign thirty $5,000 checks, checks which were subsequently voided after Kalmbach refused to help AMPI in its antitrust troubles. Additional evidence of a Mehren attempt to secure a quid pro quo from Kalmbach comes from testimony of former AMPI official Dwight Morris, who met John Butterbrodt a week later on April 11, 1972, in Chicago. Butterbrodt told Morris that AMPI representatives had an agreement with Herb Kalmbach that in exchange for AMPI paying him $300,000 ($150,000 from AMPI, and the rest from Mid-Am and DI), "the antitrust suit against AMPI would go away." Butterbrodt told Morris, however, that after the ITT scandal had hit the press, Kalmbach backed out of the deal and told AMPI that the Nixon campaign no longer wanted its contributions.[33]

In the last analysis, Mehren escaped indictment because the prosecutors had serious doubts as to whether Kalmbach's testimony would be believed by a jury:

> Of course, one can't predict these things with absolute certainty when you're dealing with such variables as demeanor, skill of defense counsel, Kalmbach's development as a witness, etc. But, it is important to point out that there is a substantial risk that Kalmbach will be discredited in several ways which would go right to the heart of the case, viz., did Mehren really say those things to Kalmbach that the government says he said. Add to this the obvious defense argument that Kalmbach's own criminal liability for various misdeeds gave him a substantial motive to embellish or fabricate the April 4 phone call testimony in order to please the prosecutors. From a defense viewpoint, this is a good reasonable doubt case because of Kalmbach.[34]

The prosecutors did believe they had substantial corroboration of Kalmbach's testimony—specifically the testimony of Bob Lilly, Harold Nelson, and Robert Isham, AMPI's former comptroller, as well as Mehren's suspicious issuance and signing of the thirty checks on April 4:

> If Kalmbach, on the one hand, and Lilly, Nelson, and Isham, on the other, hold up reasonably well, then these two incriminating facts may very well be sufficient to discredit Mehren's explanation of the phone call, indirectly corroborate Kalmbach, and thus eliminate the reasonable doubt that the defense might be able to introduce primarily through Kalmbach's crossexamination.[35]

The prosecutors candidly admitted, however, that "the case could go either way." While the reasons for their eventual decision not to seek indictment have not been made public, it is readily apparent from those portions of the prosecution memoranda which have that, while the prosecutors personally believed Kalmbach's testimony, the corroborating testimony of former AMPI employees Nelson, Isham, and Lilly, and other circumstantial evidence, they did not believe it was strong enough to convince a jury beyond a reasonable doubt.

By October of 1972, the Nixon Administration and the dairy lobby apparently believed that "the heat was off." On October 24, 1972, Lee Nunn, vice chairman of the Finance Committee to Re-Elect the President, flew to San Antonio and met George Mehren on his ranch. Three days later, Mehren had AMPI's new political action committee, C-TAPE, send a total of $352,500 in checks to the National Republican Senatorial Committee and the National Republican Congressional Committee, $221,000 of which was transferred in the next few weeks to the Nixon campaign.[36]

In July, 1973, in the midst of the Watergate scandal, AMPI's attorneys in the government antitrust case attempted to raise, for the first time, the defense that the government had brought its antitrust case against AMPI in order to extort political contributions from it. While taken off guard, Justice Department lawyers conceded that the AMPI case was brought in other than "ordinary times" and offered to open its files to prove the falsity of the AMPI charge. Before allowing such access to the Justice Department files, however, the federal judge in charge of the trial, John Oliver, required AMPI to furnish testimony under oath from those AMPI representatives who had received these alleged extortion threats. Faced with this, AMPI withdrew its extortion claim and entered into serious settlement negotiations with the government. Negotiations can be complicated in cases as comprehensive as the one against AMPI. Nevertheless, an agreement was finally reached between the parties and presented in the form of a "consent decree" to the court for its approval in August 1974. In the consent decree, AMPI essentially agreed to refrain from engaging in the predatory practices which the government alleged it had used to achieve its monopoly power. Over the objections of many of AMPI's victims, the consent decree was eventually approved and entered by the court on April 30, 1975.[37]

While no one—particularly the dairies and other farmer organizations which had suffered from AMPI's predatory and monopolistic activities—believed or implied that the government lawyers had failed to act in good faith in negotiating the consent decree with AMPI, it is also true that virtually no one other than the government and AMPI were satisfied with the consent decree. The biggest criticism was that even though AMPI had established a predatory monopoly over a large geographical area, the government's consent decree did not require or call for the dissolution or divestiture of AMPI into smaller nonmonopolistic cooperatives. Instead, it left AMPI's monopoly virtually intact. As one dairy observed in its comment to Judge Oliver on the inadequacy of the consent decree:

> AMPI achieved and maintained its monopoly position in many markets by the use of predatory practices. All the Government Consent Decree does is attempt to prohibit AMPI from maintaining its monopoly position in the future through the use of [these] predatory practices but leaves AMPI completely free to maintain its monopoly position through other means. Yet the fact remains that so long as AMPI continues to charge [excessive prices] in markets where it maintains a monopoly position, its behavior will constitute unlawful monopolization. . . .
>
> The government's myopic view of its role in the entire proceedings is revealed, albeit inadvertently, in its response to the suggestions of many third parties . . . to permit private enforcement [of the consent decree]: "The Government's statutory obligation is to prosecute antitrust violations and enforce decrees in antitrust cases in the public interest not in the exclusive narrow interest of producers, processors, retailers or consumers."
>
> Yet if one excludes producers, processors, retailers and consumers, who is there left in the "public" to protect? For that matter, if you [only] eliminate the "exclusive narrow interest" of all consumers, who is there left in the "public" to protect? The questions answer themselves. The fact is that the Government's Consent Decree ignores the public interest precisely because it ignores the interests of the public (i.e., all consumers) in being able to purchase competitively priced milk. Instead, it permits AMPI to continue to maintain illegally acquired monopoly positions from which positions it can continue to charge monopolistic prices.[38]

To many observers, it looked as if AMPI had received a wrist slap after all.

Part 4

Maintaining Power: Milk, Peanuts, and Congress

It's sad, it's cynical, but after election day fine dairy organizations not contributing to campaigns really don't have any voice or representation in Congress. . . . . What confronts us . . . is that apparently legislators won't listen to the people who elect them unless they cough up money.

John Butterbrodt, President
Associated Milk Producers, Inc., 1974

9

The Best Congress Money Can Buy

The Watergate era from 1973 to 1975 was a time of trauma for the dairy lobby, during which its excesses were exposed for all to see. More importantly, however, it was also a time of transition, a time during which the dairy lobby painfully learned the rules of the game in Washington and survived with its political power essentially intact. It survived by adapting to the tenor of the times. If illegal contributions and under-the-table payments were now in disfavor, the dairy lobby was more than willing to change its ways. After all, money was money whether it was shoveled over or under the table. The cost was the same. If Congress wanted to change other rules—as it did in the 1974 Federal Election Campaign Act —the dairy lobby would gladly comply.

Yet Congress, by passing the Federal Election Campaign Act in a moment of post-Watergate morality, should not be allowed to escape its responsibility and even culpability in nurturing and encouraging the dairy lobby during the period of its greatest excesses. For the hated Nixon and other presidential candidates were not the only politicians to benefit from the dairy lobby's questionable pre-1974 activities. They had plenty of company from Capitol Hill.

Consider Herman Talmadge. The senior senator from Georgia was first elected to the Senate in 1956. Before that, he had been a two-term governor of Georgia, having succeeded his father in that office in the time-honored tradition of Southern politicians.

He has never been seriously opposed for re-election. Running for his third term in 1968, he won with 78 percent of the vote. In 1974, he was elected with 72 percent. His campaign expenditures in 1974 were only $65,000. (Contrast this with the campaign expenditures of Georgia's junior senator, Sam Nunn, in the 1972 senatorial campaign in Georgia, when he spent $568,000, and his opponent spent $445,000). One would expect, therefore, that Talmadge would have little need for campaign contributions, particularly when his opponent in 1974 spent only $12,000. Nevertheless, Talmadge received $10,000 from the dairy lobby for his 1974 campaign—almost 20 percent of his total campaign expenditures, practically as much from one source as his opponent spent in the entire campaign. And Talmadge took the contribution, even though at the time he was serving on the Senate Watergate Committee investigating, among other things, the milk-fund scandals of the Nixon Administration. Unfortunately for Talmadge, his well-publicized and bitter divorce from his wife in 1978 exposed some of the seamier aspects of his finances from that 1974 campaign. Among other things, half of the dairy lobby's $10,000 contribution ended up in Talmadge's personal checking account[1] which in part led to his being "denounced" by the Senate after an Ethics Committee investigation.

Why would the dairy lobby spend so much money on a senator who really did not need it? And why would a senator who really did not need it accept so much from one source? The answer: Talmadge is chairman of the Senate Agriculture Committee, which must pass on all dairy legislation. A friend in that high a place is worth ten grand even though he does not need it. And because he does not need it, he will remember you with particular fondness when it is time to talk about price supports and other federal subsidies for the dairy lobby.

Another venerable recipient of dairy lobby funds on the Senate Agriculture Committee was Senator James Eastland of Mississippi, the second-ranking Democrat on the committee, now retired. Eastland had been a senator since the 1940s and had been chairman of the Senate Judiciary Committee since the 1950s. He had long been a supporter of handouts for farmers, having at times collected more than $150,000 a year himself from the Department of Agriculture not to grow cotton on his plantation in Mississippi. He understood subsidies, price supports, and other government handouts to agricultural interests. Eastland handily won re-election in 1972, and the dairy lobby came around on election day and gave him a $5,000 contribution.[2]

Being a member of the Senate Agriculture Committee can be a lucrative position. Another example is Senator Walter Huddleston (D. Ky.), who was chairman of the Senate Agriculture Committee's Milk Subcommittee, which has jurisdiction over price supports and marketing order legislation. He received $18,500 from the dairy lobby in his 1972 campaign, when he defeated former Republican Governor Louis Nunn. Huddleston was a state senator from 1965 to 1972 and since 1957 had been part owner of a small radio station in Lebanon, Kentucky. Despite his lack of an agricultural background, he asked for and received an appointment to the Senate Agriculture Committee and became chairman of its subcommmitee on Agricultural Production, Marketing, and Price Stabilization. After all, an $18,500 campaign contribution was not insignificant. The least he could do was place himself in a position of power where the favor could be repaid. What is questionable in the case of Huddleston, however, is that only $3,500 of the dairy lobby's contribution came before the election. $15,000 of the total $18,500 contribution was received *after* his election.[3]

Or consider former Senator Dick Clark (D. Iowa), who during his 1972 campaign received illegal contributions from the dairy lobby and *after* his election was given $7,500 in legitimate contributions by the dairy lobby. Clark had been a longtime Washington fixture, having worked as an administrative assistant to Iowa Congressman John Culver from 1965 through 1972. A political science professor, Clark returned from Washington to his state of Iowa long enough to get elected senator in 1972, with 56 percent of the vote. After his election, the dairy lobby trotted over, tossed in $7,500 to Clark's campaign, and Clark, as did Huddleston, asked for and received an appointment to the Senate Agriculture Committee, as well as to its Subcommittee on Agricultural Production, Marketing, and Stabilization of Prices.[4]

Analyze the pattern of contributions to these four senators— Talmadge, Eastland, Huddleston, and Clark. Talmadge and Eastland were pillars of the Senate. Neither ever had any serious opposition. They did not need campaign contributions and probably did not work very hard to solicit them. When sizeable contributions did come—$10,000 to Talmadge and $5,000 to Eastland—there could be only one purpose: access and influence. But notice that the dairy lobby does not throw money around needlessly on less than proven performers. While Huddleston was given a comparatively small $3,500 contribution prior to the election, once he won, the dairy lobby rushed to throw another $15,000 after that. The same thing happened with Dick Clark. While he was given much illegal assistance prior to the election, $7,500 of clean cash flowed into his coffers after he won.

What impels newly elected senators to accept contributions in such amounts *after* the election? If the dairy lobby genuinely believed in the merits of Messrs. Huddleston and Clark, it would have been far more generous prior to the election. The fact that the dairy lobby waited until after the election underscores its intent, and the intent of the senators in taking the money.

It paid off too. None of these four senators ever opposed increases in milk support prices. These contributions, however questionable in intent, were nonetheless legal. Other contributions to equally prominent senators were not so legal.

Senator Edmund Muskie (D. Maine) was one recipient of such illegal campaign contributions from the dairy lobby during his re-election campaign in 1970. Muskie's illegal contributions are relatively straightforward. They were laundered through attorney Stuart Russell, who admitted that he had contributed $8,400 to Muskie's re-election campaign in 1970 ($5,000 of which came after Muskie's victory), along with a cover letter saying that the contribution was being made "at the request of Bob Lilly." On the same day, Russell sent a bogus fee statement to AMPI for $7,500, precisely 150 percent greater than the post-election contribution, in order to allow Russell to pay taxes on the fees received.[5]

The receipt of illegal funds poses an interesting question for Senator Muskie. Milt Semer, Muskie's finance chief (and the courier for $100,000 in illegal cash to Nixon from the dairy lobby), knew, or should have known, who Bob Lilly was. Senator Muskie should have known as well. Both certainly knew that

Stuart Russell was an AMPI attorney. Why did this reference to Bob Lilly in Russell's cover letter not raise a red flag to Muskie or Semer? One thing we do know for certain: Muskie was one of the primary sponsors of Senate Bill 1277, introduced on March 16, 1971, to increase the milk support price to 90 percent of parity as a means of putting political pressure on Richard Nixon to reverse his earlier price support decision. Dairy lobby contributions to Senators Dick Clark and James Abourezk in 1972 were slightly more complicated than those to Ed Muskie but no less illegal.

Valentine Sherman and Associates ("VSA") was a Minneapolis-based computer-mail-service firm. It was established by Jack Valentine, a political scientist, and his friend Norman Sherman, the former press secretary to Vice-President Hubert Humphrey. The firm had a number of political clients in the 1970 congressional elections, and all of their clients were elected. The way VSA functioned was to have its employees take the names, addresses, and telephone numbers of potential voters from telephone directories and computerize the information. Once the names and addresses were fed into a data bank, additional information would be obtained through personal interviews by political volunteers trained by VSA. The additional information obtained included age, voter registration, number and ages of children, occupation, etc. If a candidate thereafter desired to send computerized mailings to, for example, all farmers in his district with college-age children, VSA would provide the names and addresses and arrange for the physical handling of the mailings.[6]

In the spring of 1971, Valentine and Sherman contacted a number of candidates in an effort to secure even more business for the 1972 elections. Many candidates complained about VSA's high prices—its standard price was $10,000 per congressional district. To overcome this problem, VSA made a pitch to various labor unions, business organizations, and the like to persuade them to subsidize VSA's potential political clients. Its approach was to ask these politically active groups to commission VSA to locate and record lists of names which they could use in their regular business. Compiling these names for the lists would then be done in conjunction with the specific needs of a potential political client of VSA. In this fashion, the politically interested organization would get a "two for one deal," i.e., a list for its use in the regular course of its business, as well as a list which would be supplied to those political clients of VSA which the politically active organization had an interest in helping.[7]

Two Washington-based, Democratic party oriented political advisers to the dairy lobby, Ted VanDyk and Bill Connell, had been favorably impressed with the work of VSA in the 1970 elections and had put Jack Valentine in contact with AMPI's David Parr. In April of 1971, Valentine arranged to fly with Parr from Louisville, Kentucky, to Minneapolis, Minnesota. While they were isolated together at 20,000 feet, Valentine made VSA's standard pitch to Parr and left with him an eight-page proposal, the gist of which was that if AMPI would pay for the gathering of the names for political purposes, VSA would furnish to AMPI the equivalent market value in lists of names and addresses demographically tailored to suit AMPI's unspecified needs. The written VSA proposal also pointed out that AMPI could win "influence" and earn "tremendous political leverage" by subsidizing VSA's political business in this manner.[8] After receiving the written proposal, negotiations continued throughout the spring of 1971 between Valentine and Sherman for VSA, and Parr and Harold Nelson for AMPI. AMPI was eventually won over and agreed to the VSA proposal.

David Parr later claimed to the Watergate Special Prosecution Force that the information gathered by VSA served three purposes. First, it allowed AMPI to make disguised contributions to Democratic candidates which would not be discovered by those in the Nixon Administration, such as Charles Colson, who might be less than favorably impressed by AMPI's "playing both sides of the street." Also, the information allowed AMPI to make an early, informed decision as to which Democratic candidate to back for president during the 1972 primaries. Finally, Parr indicated that the information *might* have been useful some day *if* AMPI ever began selling life insurance to its members by mail. This last motive, however, was clearly transparent, given Parr's admission that the primary motive of AMPI was to make "unreported contributions to candidate(s)."[9]

In 1972, then Representative John C. Culver from the second

district in Iowa, was contemplating a race against incumbent Republican Senator Jack Miller. Culver and his administrative assistant, Dick Clark, negotiated with VSA on behalf of the Iowa Democratic party for VSA's services, which were quoted to them at $60,000—$10,000 for each of the six congressional districts in Iowa. Culver and Clark wanted this survey for the state Democratic party but the party could allocate only $10,000 toward the cost of the project. Ted VanDyk persuaded AMPI to pay the remaining $50,000, an amount which allegedly would have entitled AMPI to a list of 50,000 rural names in the midwest. AMPI knew, however, that the money was going to help the Democratic candidate for senator in Iowa and deliberately paid the remaining $50,000 with corporate checks. Culver, however, subsequently decided not to enter the race. His administrative assistant and co-negotiator with VSA, Dick Clark, did and utilized the AMPI survey in his campaign. The $50,000 was, therefore, a direct illegal corporate contribution from AMPI to Clark's campaign.[10] While it became fashionable during the Watergate era for many politicians to return legal and illegal dairy lobby contributions, Clark never chose to do so. He was defeated for re-election in 1978.

In 1971, then freshman Representative James Abourezk decided to run for the U.S. Senate in South Dakota. His staff made a decision to employ VSA in the campaign but were concerned as to whether they could afford the $20,000 cost, particularly the initial payment of $7,000 which VSA wanted up front. It is unclear today whether it was Abourezk or Norman Sherman who first suggested AMPI as the source for the initial $7,000 payment. Whether Abourezk himself or Sherman was the go-between to solicit AMPI funding for the survey, the fact is that AMPI did pay by corporate check the first $7,000 payment to VSA on October 18, 1971, and it was an illegal corporate campaign contribution. Abourezk, like Clark, never returned the contribution. He did not run for re-election.[11]

Sherman and Valentine told the Watergate Special Prosecution Force that when VSA and AMPI first established their relationship, they assumed all bills would be paid with TAPE funds. When the first AMPI corporate checks were received by VSA in July of 1971, Sherman claims he realized the implications of receiving corporate money. He nevertheless deposited the check and sought the advice of none other than Minneapolis attorney and Hubert Humphrey campaign manager, Jack Chestnut, who told Sherman that so long as AMPI was receiving some tangible benefit (i.e., lists of names) for its corporate payments, the transaction was legal. Chestnut thereafter drafted a contract to cover the arrangement. Part of Sherman's eagerness to cash the first AMPI check was undoubtedly based on the fact that at the time the first AMPI check arrived, VSA was overdrawn at the bank and needed the money.[12]

That VSA and AMPI felt guilty about their questionable relationship is demonstrated by what the Watergate Special Prosecution Force termed, "the next phase of the conspiracy . . . involv-[ing] the fabrication of backdated correspondence and invoices to disguise the true nature of the transactions." Bob Lilly flew to Minneapolis in December of 1971 in AMPI's corporate jet and visited VSA's offices. At that time, according to Lilly, Valentine expressed concern over the lack of correspondence between VSA and AMPI and stated he wanted to prepare a correspondence file to "back up" the billings. Lilly, with the consent of Harold Nelson, then sent blank AMPI letterhead to Valentine, who prepared the false documents for Lilly's signature. In March of 1972, Valentine called Lilly and told him that the correspondence and false duplicate invoices had beeen prepared. On March 23, 1972, Lilly took a commercial flight to Minneapolis and met Valentine in the Braniff Club at the Minneapolis airport and signed all the letters. Valentine then gave Lilly copies of the fabricated letters as well as false duplicate invoices.[13]

Valentine does not deny that all this happened. He merely claims that it was all Lilly's idea and that Lilly asked him to create the new invoices and correspondence and furnished VSA with blank AMPI stationery for that purpose.

At approximately the same time that the fabricated documents were signed by Lilly, he asked VSA to furnish AMPI with some of the names it had on file. VSA accordingly sent six reels of computer tape containing approximately one million names. Upon receiving the computer tapes, however, AMPI put them in storage and never made use of them. The reels were sealed at the time of delivery to AMPI in 1972 and the seals never broken until the internal investigation of AMPI conducted for its Board of Directors by the Little Rock, Arkansas law firm of Wright, Lindsey and Jennings in the latter part of 1973 and early part of 1974. The six reels of tape which were delivered to AMPI contained names, addresses, zip codes, and telephone numbers of individuals residing in the states of Iowa, South Dakota, Kansas, Oklahoma, Nebraska, and Minnesota. According to VSA's attorney, the total amount of money paid by AMPI to VSA—$137,000—would have entitled AMPI to approximately five and a half million names and addresses on the basis of VSA's usual rate. AMPI, of course, never received all five and a half million names and never used the one million it did receive. VSA had at that time only twenty million names and addresses on computer tapes, which means that AMPI had paid for over 25 percent of VSA's data banks.[14]

Another politician who did make use of VSA's data banks in the 1972 campaign was Vice President Walter Mondale, a man familiar with the dairy lobby and its money. Then Minnestoa's senior senator, Mondale had been co-chairman of Hubert Humphrey's 1968 presidential campaign, which received at least $32,000 in illegal campaign contributions from AMPI's predecessor, MPI. In 1971, when Agriculture Secretary Hardin first announced that the Department of Agriculture would not raise the milk support price to 85 percent of parity, Mondale was one of the primary sponsors of legislation to overturn Hardin's original decision. Although the legislation proved unnecessary when Nixon reversed Hardin's decision, the dairy lobby was not ungrateful.

In late winter of 1971, some time in February or March, Senator Mondale had a breakfast meeting in Washington, D.C. with Harold Nelson and Dave Parr. At that meeting, Nelson and Parr promised the dairy lobby's support to Mondale's 1972 re-election campaign. Parr later solidified that commitment by telling Mondale and Ted VanDyk that AMPI would raise $25,000 for a Mondale fund-raising dinner scheduled for June 5, 1971 in Minneapolis.[15]

Michael Berman, Mondale's campaign manager for his 1972 re-election campaign, called Bob Lilly on April 28, 1971, and asked him to go to Minneapolis to help sponsor, work for, and raise $25,000 for the June 5 fund-raiser. Berman told Lilly that Dave Parr had promised Mondale *both* the $25,000 and the personal assistance of Lilly. Lilly said he never heard of the promise.[16]

Berman called Lilly several times during the next month regarding the $25,000 and finally, on May 27, 1971, Lilly called Harold Nelson for instructions. Nelson told Lilly that he was unaware of any such commitment but he would check with Dave Parr and get back to Lilly. Nelson later called Lilly back, saying he could not reach Parr but would get in touch with Lilly when he did. Nelson apparently never did call Lilly about this again, and Lilly, in the meantime, did nothing.[17]

Ted VanDyk was the next person to contact Lilly about the $25,000 commitment to Mondale. VanDyk told Lilly the details of Mondale's Washington breakfast meeting with Nelson and Parr and Parr's promise to raise $25,000 for the June 5 fund-raising dinner. VanDyk then gave Lilly names and addresses for three committees to which the contributions for Mondale should be sent and instructed Lilly to call Mike Berman and bring him up to date. Lilly declined.[18]

AMPI's political fund finally did contribute at least $5,000 to Mondale's June 5 fund-raising dinner. Mondale, as the laws then allowed, never had to make a full public accounting of his receipts from that dinner. So he didn't.[19] About one month after Mondale's June 5, 1971 fund-raiser, AMPI sent a corporate check for $25,000 to VSA. The $25,000 was part of the $137,000 paid by AMPI to VSA for computerized mailing lists in six midwestern states, *including* Minnesota. Lilly's handwritten notes contain the following reference regarding this $25,000 payment: "To Mondale? To HHH?"[20]

Did Mondale benefit from any of this? No one can be certain. VSA *did* do work for Mondale in the 1972 campaign; AMPI *did* pay for compiling approximately six million names and addresses, over 25 percent of VSA's stockpile of names; and Mondale *did* pay for VSA's services with either his own campaign committee's or his party's campaign funds. The question is, however, what Mondale paid for. Was VSA compensated by Mondale for its costs in compiling and storing the Minnesota names and addresses in the computer banks? Or did Mondale simply pay for: (1) the cost of retrieving the names already in the computer and (2) mailing the literature? It makes a difference because AMPI paid VSA for compiling names in Minnesota. If Mondale didn't pay for compiling names, but simply used names and addresses compiled at AMPI's expense, then Mondale received an illegal campaign contribution just like Humphrey. VSA knows the answer but is not telling, and its reconstructed records for this era are of scant probative value.

The House of Representatives has been just as susceptible to dairy lobby cash as the Senate. The only difference is that there are more representatives than senators. Individual contributions to representatives have usually been smaller than those to senators. Page Belcher was an exception to this. Nevertheless, the story of his illegal support illustrates the symbiotic relationship among Congress, the Department of Agriculture, and the dairy lobby.

Page Belcher was Tulsa's congressman from 1950 through 1972, when he retired from Congress. He had been the ranking minority member of the House Agriculture Committee and, prior to 1970, had not had a close re-election in the previous twelve years. In 1970, however, Belcher found himself in a tough fight.

During his years in Congress, Belcher had always been a strong supporter of high milk price supports and high dairy import quotas. It was only natural, therefore, for Belcher to look to the dairy lobby for help when he found himself in trouble in 1970. While all the details are not known—most of them being shrouded in grand jury secrecy—the internal memoranda of the Watergate Special Prosecution Force indicate that Belcher received $20,000 in illegal corporate political contributions from AMPI during that campaign.[21]

After his tough campaign in 1970, Belcher decided he had had enough of electoral politics and retired at the end of his term. With the Nixon Administration in power, however, Belcher saw no reason to leave politics altogether, particularly given his long years of experience on the House Agriculture Committee, where he had generally worked hand in glove with the majority members. Besides, Belcher had friends in high places. On June 28, 1972, Nixon's secretary of agriculture, Earl Butz, sent a letter on official

USDA letterhead addressed to George Mehren, general manager of AMPI. Butz bluntly urged AMPI to take care of his friend Page:

> [AMPI] may have an interest in making use of Page's many talents and broad contacts on the Hill. . . .
>
> I know that we can make good use of Page next year as we undertake to hammer out a new agricultural bill, or an extension of the present bill.[22]

"Dear Earl," wrote "George," on July 5:

> I thank you most sincerely for your letter relevant to the plans by Mr. Page Belcher. . . . You will be glad to know that we have been in contact with him and will avail of his services.[23]

Belcher's services turned out to be worth $75,000 a year to the dairy lobby, including $15,000 worth of expensive office furnishings befitting his new status as a lobbyist. Unfortunately, the *Wall Street Journal* broke the story in September, 1973, revealing the exchange of letters between Butz and Mehren, and Butz's role as shill for Belcher in setting him up with the lucrative lobbying position. The parties to the triad were not pleased with the revelation. Butz was particularly unhappy. If Butz could not use his official position to find lobbying positions for his friends (positions which could have a direct effect on USDA policy), then "to hell with it . . . that may not be the way other people do it, but it's the way Earl Butz does it."[24]

Belcher's retirement from Congress, however, did not loosen the dairy lobby's grip on Oklahoma's first district congressional seat. Belcher was succeeded by James R. Jones, who had been on President Johnson's White House staff, and had served as LBJ's appointments secretary in 1968. Jones was no stranger to the dairy lobby. He first met AMPI's Harold Nelson and David Parr while working at the White House. In 1969, shortly after leaving the White House, Jones returned to Tulsa and promptly started campaigning against Page Belcher, winning 44 percent of the vote in 1970. In 1972, he won Belcher's seat with 55 percent of the vote. Jones ran well ahead of the national Democratic ticket, George McGovern receiving only 21 percent of the vote in the district.[25]

Jones was not without his own dairy skeletons. Before his election to Congress, Jones had written a memorandum to AMPI, claiming credit for a favorable price support decision made in the latter days of the Johnson Administration after Nixon's election and openly soliciting legal business from AMPI as a reward. The dairy lobby does not forget its friends and, as a result, Jones was given a $40,0000 annual legal retainer by AMPI. Jones had a further arrangement with AMPI, whereby he billed it for additional compensation at the end of its fiscal year, if the amount of his legal services rendered exceeded the amount of the retainer. Jones also served during this period as editor of *Dairymen's Digest,* an AMPI publication.[26]

With uncanny timing, Jones's memorandum soliciting legal business from AMPI was made public by the Senate Watergate Committee on the same day AMPI pleaded guilty and was fined $35,000 for, among other things, making illegal campaign contributions to the 1970 campaign of Jones's predecessor, Page Belcher. Jones, understandably displeased with the revelation of his ethically questionable behavior, claimed that he was "sick, and angry with medication," when he wrote the memo and that he had never intended it to be mailed to AMPI (even though it had been).[27]

While it was never given much publicity, Jones's position as one of AMPI's outside lawyers led to his direct involvement as a conduit in connection with the complicated machinations surrounding the dairy lobby's original illegal payment of $100,000 in cash to Herbert Kalmbach in August, 1969. Jones was one of the AMPI lawyers who was asked to, and did, contribute $10,000 directly to Bob Lilly. The purpose was to enable Lilly to repay the $100,000 bank loan he had personally taken out to reimburse AMPI's political fund, which had never reported the $100,000 given to Kalmbach. The way the scheme worked, according to the Watergate Special Prosecution Force, was that "the attorneys could 'make themselves whole' by inflating their billings to AMPI. The loan was paid off over the course of the next year by utilizing funds given to Lilly by the attorneys. . . ."[28]

Jones denied that he ever inflated his billings to AMPI as part of their scheme. Initially, he even denied giving the money to Lilly, claiming to the Senate Watergate Committee that he had given the money to TAPE, AMPI's political action committee. He changed that story when confronted with his own cancelled checks.

Jones stuck by his claim that he had never inflated his billings to AMPI to make himself whole for the $10,000 he gave to Lilly. This story, however, has a hollow ring to it. Jones had given Lilly $5,000 on two occasions. Each time he gave money to Lilly, he did so only *after* he had submitted a *special* billing to AMPI approximately 40 percent greater than the $5,000 he gave to Lilly— $6,890 on one occasion and $7,150 on the other. In the three years Jones worked for AMPI, he received a monthly legal retainer and expenses. These two special billings were the *only* other legal fee statements he submitted to AMPI over that three-year period. Moreover, Jones never explained what, if any, special projects he performed for AMPI to warrant such special billings. AMPI's general manager, Harold Nelson, could not recall any special work performed by Jones either.[29]

AMPI's subsequent actions confirmed that it believed it had been overbilled by Jones as part of the Lilly reimbursement scheme. In January 1974, AMPI requested the Finance Committee to Re-Elect the President to refund the $100,000 contribution given to Kalmbach in 1969, because of its corporate source. Since the $10,000 from Jones was used to reimburse TAPE via Lilly for its payment to Kalmbach, AMPI could hardly claim a corporate source for the $100,000 to Kalmbach, if it had not, in fact, reimbursed Jones through overbillings. The Watergate Special Prosecution Force believed the same thing. Its memorandum on Harold Nelson comments generally upon conduits like Congressman Jones and states that "it is sufficient to note that Lilly's repayment of the $100,000 contribution to TAPE is an "indirect payment" of corporate funds since *the ultimate source of the funds is AMPI rather than Lilly or the attorneys*" (emphasis added).[30]

Congressman Jones was never indicted for his role in the AMPI-Lilly-TAPE reimbursement conspiracy; however, he did not escape from Watergate unscathed. In January 1976, he pleaded guilty to a misdemeanor for failing to report an illegal 1972 campaign contribution from Gulf Oil. It did not affect his popularity with the voters, however, as he carried his district in 1976 with 54 percent of the vote while Jimmy Carter was losing there with only 38 percent.

Today, Jones is still in Congress voting for higher milk price supports and higher dairy import quotas whenever he gets the chance.

In 1974, during the white heat of Watergate, the dairy lobby noticeably lowered its political profile. It limited its campaign contributions to a total of $102,970 distributed among thirty-five congressmen during the 1974 campaign, all of whom subsequently voted right after the election (on December 20, 1974) to raise milk support prices. By this time, however, the dairy lobby had become even more sophisticated in the timing of its contributions; $67,670 of its total in 1974 was contributed after October 24, 1974, the closing date for the last public report to be made prior to the election. Because the lobby made the contributions *after* that date, voters had no chance to learn of it until the election was over. Mid-Am was particularly clever about its last-minute donations. Six of them, consisting of $4,995 each, were made beginning on October 25, 1974.[31] This tactic neatly avoided the additional legal requirement that donations of $5,000 or more made in the last few days before an election be reported by telegram. Had Mid-Am's donations been a day earlier or had they been $5 more, voters would have had a chance to learn of the donations before the election.

Those congressmen who received the last-minute donations were Representative Theodore M. Reisenhoover, a Democrat from Oklahoma's Second District and owner and publisher of the *Tahlequah, Oklahoma Pictorial Press* and *Star Citizen*; Representative Bob Traxler of Michigan's Eighth District, who had earlier been elected that April in a special election and had been appointed to the Agriculture Subcommittee of the House Appropriations Committee; Representative Bill D. Burlison, Democrat from Missouri's Tenth District, who was, like Traxler, a member of the Agriculture Subcommittee of the House Appropriations Committee; and Representative John P. Murtha, Democrat from Pennsylvania's Twelfth Congressional District and also a member of the House Appropriations Committee.[32]

Two Republicans, William J. Scherle of Iowa and Robert Mathias of California, received last-minute donations from Mid-Am and both lost. Mathias had been the sixth-ranking Republican on the House Agriculture Committee, and Scherle had been on the Agriculture Subcommittee of the House Appropriations Committee. Scherle had been one of the few Republicans in March of 1971 to introduce a bill to increase the parity support price of milk to between 85 and 90 percent.[33]

Not all congressmen were pleased to receive campaign contributions from the dairy lobby in 1974. Or, rather, after receiving them without objection, some congressmen suddenly decided that they were becoming more of an embarrassing liability than an asset. Accordingly, at one point in 1974, there was a widespread attempt by various and sundry politicians to disassociate themselves from the dairy lobby by publicly returning its contributions. Among those who publicly returned contributions were Senators Robert Dole of Kansas, who returned almost $16,000; Adlai Stevenson of Illinois, who returned $1,500; Gaylord Nelson of Wisconsin, who returned $250; and Charles Mathias of Maryland, who returned $5,000.[34]

Representatives who returned dairy lobby contributions include Abner Mikva of Illinois, who returned $2,000; Thomas Railsback, also of Illinois, who returned $500; Charles Rangle of New York, who returned $100; Wayne Owens of Utah, who returned $100; Frank Denholm of South Dakota, who returned $5,000; and Richard Nolan of Minnesota, who likewise returned $5,000.[35]

By 1976, the dairy lobby was back in the campaign contribution business in a big way, its role having been validated by Congress itself in the 1974 Federal Election Campaign Act. Mary Meehan, writing in *Inquiry* magazine, has characterized that supposed "reform":

> . . . The Federal Election Campaign Act . . . cannot be understood at all if viewed as an altruistic reform. The key to understanding the way it was written and the way it works is to view it as a treaty among five special interests: the Democratic party, the Republican party, incumbents, big labor and big business.
>
> After a war called Watergate, the great powers met to hammer

out a treaty. The task of the 1974 peace conference which took place in Congress with assistance from the outside, was to divide financial territory for the post-Watergate era . . . in such a way as to give an appearance of integrity and fairness. And if the treaty itself, like the Versailles Treaty after World War I, did not have that appearance, then opinion leaders must be persuaded to make the public think that it did.[36]

There were several aspects to the "reform" which were most congenial to the dairy lobby and other special business-interest groups and to all the incumbent congressmen with whom they share a symbiotic relationship. *First,* the reform imposed limits on the campaign spending of all congressional candidates. This was crucial to incumbents because if their challengers could not spend more than they, these challengers rarely could defeat them. The dairy lobby and other interest groups liked this limitation because their contributions would carry more weight with less money than if there were no limits. The Supreme Court later struck this provision down.

Second, the reform imposed limits on the contributions of individuals to any candidate for Congress or president—$1,000 for primaries and $1,000 for the general election. If the dairy lobby liked the first reform, it loved this one. By an act of Congress, the dairy lobby and other special interest groups had been openly declared to be more important than individuals. Five times as important, to be precise, because the spending limit for political action committees was $5,000 for primaries and another $5,000 for general elections.

The symbolism of this disparity between the treatment by Congress of individual citizens and special interest groups should not be underestimated. Organized pressure group politics is the way Congress has worked—witness the example of the dairy lobby—for at least the last forty years. The 1974 Federal Election Campaign Act merely institutionalized the contempt in which Congress holds individuals and the high regard they have for special interest groups. For those who still have any doubts that this is so, the dairy lobby's role in the 1976 elections and the subsequent payoff by Congress in less than a year should help remove them.

From its Watergate era low profile of only $100,000 in contributions to congressmen in 1974, the dairy lobby returned in 1976 literally carrying baskets of money: almost $1.4 million spent by the dairy lobby to elect a 95th Congress congenial to its interests. An analysis of where this money was distributed demonstrates why the dairy lobby is today—in sheer terms of achieving in Congress its narrow economic interests which benefit its dairy farmer constituents at the expense of all other consumers—the most spectacularly successful lobby in this century:

- $89,336 given to eight of the eleven members of the Dairy Subcommittee of the House Agriculture Committee, almost as much to one small, crucial subcommittee as it gave to the entire Congress two years earlier.
- $191,276 given to twenty-seven of the forty-six members of the entire House Agriculture Committee, the one which originally considers and holds hearings on all legislation directly affecting the economic interests of dairy farmers.
- $67,550 given to nine of the twelve members of the Agriculture Subcommittee of the House Appropriations Committee, the one which initially passes on and authorizes appropriations which the dairy lobby eases through Congress—like milk price support increases.

Do campaign contributions like these really influence votes in Congress? The beneficiaries deny it. Money can not buy their votes, no sir. The donors have a slightly different opinion. John Butterbrodt, AMPI's former president, once elaborated on the dairy lobby's view of Congress and campaign contributions:

> It's sad, it's cynical, but after election day fine dairy organizations not contributing to political campaigns really don't have any voice or representation in Congress. We dairy farmers and dairy industry leaders can talk here all we want about what should be done but unless Congress listens, nothing happens. What confronts us . . . is that apparently legislators won't listen to the people who elect them unless they cough up money.[37]

The Congress elected in 1976 had no hearing problems whatsoever when the dairy lobby's voice was raised. The big payoff came when Congress obligingly passed the Food and Agriculture Act of 1977, otherwise known as the "Omnibus Farm Bill." The bill was not particularly lavish as these measures usually go and hence was unpopular with farmers generally, particularly wheat and grain farmers, who had wanted guaranteed prices equivalent to their cost of production. Farm state congressmen criticized the bill as "the worst farm program in twenty-three years" and "cruel and unusual punishment for wheat producers."

The dairy lobby had no such complaints over the Omnibus Farm Bill and walked away from Capitol Hill wearing a big, wide grin. Its $1.4 million in campaign contributions for 1976 had just been rewarded by a guarantee of an *increase* in direct government expenditures for the milk price support program of nearly $4 billion over the next five years.[38] The rate of return alone on that investment by the dairy lobby is simply staggering.

Who made this investment by the dairy lobby pay off so handsomely? The same senators and representatives who took the campaign money and solemnly assured every reporter who would listen that no one could buy their vote: All but two of the members of the House Agriculture Committee who received dairy lobby cash voted for the $4 billion windfall. All of the members of the Agriculture Subcommittee of the House Appropriations Committee who received dairy lobby cash voted for the bill.

It was, in plain terms, one of the biggest rip-offs of government funds ever perpetuated by a single vested economic interest. And that was only for starters. Congress guaranteed the dairy lobby $4 billion dollars in increased *direct* government expenditures over the next five years. The dairy lobby was still free to go out in the marketplace and use that guarantee as leverage to gouge even more from consumers.

By the spring of 1979, the USDA reported that price increases on dairy products sold to consumers were rising by $3 billion a year, which would mean $15 billion if the same rate continued over five years. Four billion dollars received from the government and fifteen billion dollars taken from consumers. If only Congress had been a little more hard of hearing.

10

Carter Country: Peanuts, Ice Cream, and Cows

Jimmy Carter is not hard of hearing either. What was he doing about all this when his own Agriculture Department announced in May, 1979, that milk prices were increasing at a rate of 12 percent a year and that milk cost consumers $3 billion more than the year before?

Writing letters. Carter had his aides on the President's Council on Wage and Price Stability write several hundred letters to milk processors across the country asking them for information on their milk prices as a "first step" in finding out why they have been rising so rapidly. Jimmy Carter's own Agriculture Department works hand in glove with the super-cooperatives of the dairy lobby in fixing raw milk prices at an artificially high level, which milk processors by law are required to pay, and Carter's "first step" was to ask the victims why prices were so high. It is a performance worthy of the response the lawyer and critic Huntington Cairns once made to a reporter who obtusely refused to agree with one of Cairns' more obvious opinions: "If you are blind, how can I make you see the color blue?"[1]

Blindness is the best you can say of Jimmy Carter and his policies toward the dairy lobby. It may be too kind. The evidence suggests that Jimmy Carter is not really blind when it comes to the interests of the dairy lobby. He has been there. As a result, the dairy lobby likes Jimmy Carter. They think he is swell. They especially like his open style. Carter and the Democratic National

Committee openly received over $126,000 in campaign contributions from the dairy lobby during the 1976 campaign. Carter then bought himself an uphill victory against Gerald Ford in Wisconsin by promising that state's dairy farmers to raise the government's support price for surplus milk. Less than three months into office, Carter openly presented American consumers with the bill for his Wisconsin victory—$1.2 billion in increased retail milk prices, a foretaste of the bigger rip-off which Congress was to pass later that year in the form of the 1977 Omnibus Farm Bill.

The bill collector was Robert Selmer Bergland, Carter's secretary of agriculture. Like his boss, Bergland had a down home aversion to his given names and was known to all as plain "Bob." Described by editorialists as a rabid profarmer, Bergland had indeed been a farmer as well as a minor bureaucrat in the Department of Agriculture. He was also a frustrated politician until his 1970 election to Congress from Minnesota's Seventh District. In the House, he labored as a junior member of the Agriculture Committee. He remained there until after the 1976 presidential election, when he was rescued from congressional obscurity at the urging of Vice President Walter Mondale.

On the job for a mere three months, Bergland had to face a hostile press conference and deliver the less than pleasant news that the Carter Administration was going to gouge approximately $1.2 billion from consumers (by an increase in the milk price support level) and safely deliver it to the dairy lobby to pay off a Carter campaign promise. Bergland's job was not made easier by the president's own Council on Wage and Price Stability which, earlier in March, 1977, advised Carter that the current milk support price (at 75 percent of parity) was too high. Any increase, the council reported, will be "of benefit to no one and can best be viewed as pure waste."[2]

On the other hand, the political allies of Carter and Bergland had not been helpful either in paving the way for this latest in a long line of consumer ripoffs by the Department of Agriculture. Representatives of the dairy lobby had publicly announced that an 80 percent parity level was the largest increase they thought the market could stand at that time.

Carter thought otherwise. He was not going to be the first

elected president in forty years to let market considerations override politics when it came to setting a milk support price. While
Bergland echoed the dairy lobby and urged Carter to set the price
at 80 percent of parity, Carter, faithful to his Wisconsin campaign
promise, wanted to make it 85 percent of parity. Following his
own Wage and Price Stability Council's advice was never even
considered. Bergland and Carter eventually compromised and decided to increase the milk support price to a then unprecedented
$9.00 per hundred pounds—slightly over 83 percent of parity.
This would cost consumers over $1 billion a year but was "one of
those things that is considered acceptable," according to a Carter
deputy assistant for policy analysis.[3]

At his press conference, Bergland came forward to face the reporters, promptly got the bad news out, admitted it was a political
payoff, and then offered a plethora of platitudes: the gouge was
"necessary due to the very high costs of fodder and forage which
had been brought about as a consequence of the drought"; and
without a "modest price support increase," there would be heavy
attrition among dairy farmers, resulting in "extremely high prices"
within a year or two.[4] Bergland cited no studies, reports, fore-
casts, or even guesses from any authoritative sources to support
his questionable economic conclusions. In fact, available Department of Agriculture research at the time flatly contradicted Bergland, i.e., the Department of Agriculture had predicted normal
weather conditions and an *increase* in milk production without a
price support increase.

How, one reporter challenged Bergland, was he going to justify
the inflationary impact of the price increase to the Council on
Wage and Price Stability? "Well," snapped Bergland, "I am not
going to justify it to them." Bergland then tentatively advanced the
rest of his story: the increased price supports would not affect the
cost of milk to the consumer ("I can't imagine that there should
be any substantial or noticeable increase of cost of milk to house-
wives as a result of this").[5]

Bergland was not a dairy farmer himself. Clearly beyond his
depth, he had brought with him to the press conference Sidney
Cohen, chief of the Dairy Products Branch of the Department of
Agriculture's Commodity Operations. At this point, Bergland

referred further questions to Cohen. Cohen refused to support Bergland's explanation and specifically stated that "there could be some impact at retail levels."

The press smelled blood. Another question to Cohen: "It seems to me that I misunderstood. The secretary said that the increase would not be in the fluid milk area and yet you are saying that there will be an increase." Bergland interrupted before Cohen could answer. "Maybe I am the one that made the mistake," said Bergland, trying to hold on to his rapidly fading credibility. "Let me ask," he said to Cohen, "is there a direct pass-through, *penny for penny,* in the fluid milk market even though this deals only with the manufactured milk market?" Cohen, a respected dairy economist, was also an experienced bureaucrat. He had been with the Department of Agriculture back in March, 1971, when Richard Nixon made a similar political payoff to the dairy lobby. He had not then and did not now prostitute his reputation on the altar of politics: "There is a more direct relationship than on the surface, sir," he politely advised Bergland.

Upstaged by his aide, Bergland turned to the reporters, pleaded ignorance, and lamely admitted that there was a more direct relationship than he had thought. The press continued its interrogation of Cohen. "The consumers will notice a difference?" he was asked. "Most likely," Cohen replied. "If it is six cents per gallon wholesale, is it likely to come through at roughly that amount retail?" one reporter asked. Cohen agreed that this was correct.[6]

Clearly Bergland's decision was *entirely* political. There were no public policy considerations, nothing but a Carter political payoff. A payoff akin to Richard Nixon's 1971 milk price support decision, which was prompted by the same two considerations: campaign cash and political support. Indeed, Bergland had cared so little for the economic and public policy considerations involved that he never took the time to be briefed by his staff on the basic details of how milk price supports operate. Highly placed Department of Agriculture officials readily conceded that Bergland's decision was based *entirely* on Carter's campaign promise and nothing more.

Illegal? You bet it is. Clifford Hardin, Nixon's agriculture secretary, had been under investigation by the Watergate Special Prosecution Force because of his similar role in Nixon's 1971 price support decision. Cases like this involving the state of mind of a government decision-maker are difficult to prove (as the Watergate Special Prosecution Force admitted in an internal memorandum on Hardin), and Bergland was fortunate in that, unlike Hardin, he never had to swear under oath regarding his actions and place himself in criminal jeopardy. Nevertheless, criminal liability aside, the law is clear that milk price support decisions *must* be based on statutory criteria alone. Political campaign promises and gratitude for campaign contributions do not number among those criteria.

Public reaction to the Carter milk price support payoff was hostile. Ralph Nader characterized it as "an enormous gouge to the consumer" which "cannot be economically justified on the basis of legitimate concern for milk producers."[7] An editorialist from the *Cleveland Press* had similar comments:

> The next time the Carter Administration talks about "fighting inflation," don't pay any attention. Such talk is meant merely to gull suckers, not to be taken seriously by intelligent people.
> This is the grim conclusion one must draw from Agriculture Secretary Bob Bergland's unwarranted move—approved by President Carter—to boost milk price supports to $9.00 per one hundred pounds from the present $8.26.
> Bergland mumbled something about hoping that the new level will prevent milk shortages and even higher prices in the future, but that was self-serving nonsense.
> The real story is that Candidate Jimmy . . . promised the sky to dairy farmers, and now Bergland is paying off.[8]

The *New Republic* carried an analysis of the Nixon and Carter price support decisions titled "Two Milk Duds," with the subtitle "Carter Does It for Free." Comparing Carter's campaign promises regarding inflation, and his promises to Wisconsin dairy farmers, the article commented:

> It is difficult being a man of your word when you make conflicting commitments. In this case, Carter's promise to the public that he would do his best to hold down inflation was the one he broke. . . . One explanation, driven home by the Nixon and Carter episodes, is that all the fancy calculus in the end has little to do with how the support price is set. The price of milk appears to

be a political decision. Bergland is from a dairy state, milk co-operatives have been large contributors to his past congressional campaigns, Carter was grateful to the dairymen so a "fair" milk support price is established. You don't have to be bribed to want to take care of your own.[9]

Barron's put it even more succinctly:

The science of politics advanced by leaps and bounds last month when Bob Bergland, secretary of agriculture, admitted that the price of milk was going up because Jimmy Carter had made a promise to the voters of Wisconsin. Hitherto, political science was an inexact discipline, but the Bergland disclosure makes possible a remarkable new precision. Dividing Wisconsin's share of the milk-price boost this year (which runs to $167 million) by the number of its electoral votes, eleven, yields $15.2 million. This is the price per electoral vote. It may seem steep to the taxpayers but it comes virtually free of charge to President Carter and the Democratic National Committee. Nor does it offend the Federal Elections Commission.[10]

By January 1978, the adverse effects of the Carter-Bergland price support payoff was so great—i.e., the resulting milk surplus from dairy farmers responding to the high prices was so large—that both the dairy lobby and the Department of Agriculture were frantically searching for any way to reduce the surplus other than decreasing the support price for milk. One such proposal came from Congressman Jim Jeffords (R. Vermont). Jeffords' bill offered dairy farmers increased financial incentive to "cull" (i.e., reduce) their herds. The bill would have increased the "cull cow" price by about fifteen cents a pound for a certain period. This would hopefully cull approximately 500,000 cows, thereby reducing milk production by three billion pounds. Jeffords proposed specific requirements for farmers to participate in the program: voluntary production restrictions for six months to a year; proof of the number of cows owned prior to the effective date of the law; and proof of the number owned after their sale to the government at the special incentive price.

Jeffords managed to get Representative Charlie Rose (D. North Carolina) to hold hearings in April, 1978, on the huge milk surplus created by the Carter-Bergland payoff, using Jeffords' "Dairy

Herd Reduction Act of 1978" as a focus. Department of Agriculture witnesses testifying at the Rose hearings told him that they would like to have more flexibility in setting price support levels as a means of dealing with the surplus problem, i.e., setting it at 75 percent of parity in order to eliminate the surplus, something the Council on Wage and Price Stability had been saying a year earlier. Rose was having none of that. All Rose wanted was to be convinced that the Department of Agriculture was "doing everything it possibly can to creatively find ways to move the dairy product surpluses" about which Rose was "very concerned."[11]

As for Jeffords' Dairy Herd Reduction Act, the dairy lobby did not really like it and neither did the Department of Agriculture. The dairy lobby disliked it because of the controls necessary to make sure that the Act was properly carried out. The Department of Agriculture opposed it because they calculated that of the 330,000 cows culled under the proposed act, 300,000 of them "would be replaced immediately," and that at the end of the year, "both milk production and cow numbers would return to their normal levels."[12]

What the dairy lobby (through The National Milk Producers Federation) advanced as an alternative was the ploy of requesting the Department of Agriculture to increase its existing dairy indemnity program whereby the Department of Agriculture actually pays dairy farmers to depopulate or cull herds chronically infected with brucellosis and tuberculosis. The dairy lobby claimed that if the Department of Agriculture asked Congress for a supplemental appropriation to increase the dairy indemnity program by $50 to $100 a head, this would help decrease the huge milk surplus, i.e., pay the farmers to kill off sick but otherwise producing cows.[13]

This "diseased cow" proposal illustrates the bankruptcy of the government's milk price support program. The supports are so high, so out of line with reality and the market, that in order to gain the largess being offered by the Department of Agriculture, dairy farmers will milk any cow, sick or healthy, in order to participate. It also shows how the many effects of government intervention cannot be predicted. When the government intervenes to establish a higher than justified price for milk, gouging the consumer with one hand, it thereby creates another, heretofore unforeseen, problem—increasing the production of milk from infected cows. What is the dairy lobby's solution? Does it want to eliminate or drastically reduce the price supports which caused the increased production from infected cows in the first place? Not on your life. Instead, it proposes to spend more government money to *persuade* the dairy farmers with infected cows to sell them to the government. Why have one government subsidy when you can have two, particularly where the second is caused by the first.

The Department of Agriculture's desperation about the milk surplus was illuminated further by Carol Foreman, assistant secretary of agriculture for Consumer Services, who announced in January, 1978, that schools participating in the National School Lunch Program had been invited to take dairy products at no cost. This invitation, according to Foreman, would continue so long as the dairy products remained in abundant supply (a safe bet) and as long as they were being properly used "in the operation of both the school lunch and price support programs." It was a candid admission by the Department of Agriculture that if it could not give dairy products away to schools, it would not know what to do with the huge surplus created by the Carter-Bergland price support payoff.[14]

Early in his administration, Carter had momentarily thrown the dairy lobby into a panic when he proposed appointing Carol Tucker Foreman as an assistant secretary of agriculture. On the surface, the dairy lobby had good reason to be worried. Foreman had been executive director of the Consumer Federation of America. With some thirty million members, it was the largest consumer group in the country. Appropriately, Foreman was the highest paid consumer lobbyist in Washington and, by all accounts, one of the best. Bright, knowledgeable and articulate, she knew what she was doing. Under her leadership, the Consumer Federation of America unanimously passed a resolution attacking the antitrust immunity of agricultural cooperatives, particularly those (like the large dairy cooperatives) possessing monopoly power.

And she did not accord USDA the sacred cow status to which it had long thought itself entitled. In one speech, she attacked both the USDA and the dairy lobby:

... The Capper-Volstead Act directs the secretary of agriculture to ... hold a hearing and issue a cease and desist order if he has reason to believe that the cooperative has monopoly power. ... To my knowledge, the secretary has never acted under this power. I think it's time somebody filed a suit to order the secretary to carry out his functions as mandated by the Congress. Clearly the United States Department of Agriculture is not going to rush out and do this job, and let me assure you that if they did rush out and do a job, most of the consumers would not like it. The department is not notorious for its high standing and credibility among American consumers.[15]

Despite her reservations about the Department of Agriculture, Foreman accepted the appointment; Carter made the appropriate proconsumer noises after Foreman's Senate confirmation "an innovative thing for the U.S. Department of Agriculture ... she has complete freedom and direct instructions from me to be a strong and able advocate of consumer progress in the future"; and the dairy lobby breathed a collective sigh of relief.[16]

Why? Because Carter had given Foreman only half a job. The position she filled had in past administrations been assistant secretary for marketing and consumer services. As such, it was a powerful one, because it had jurisdiction over all of the Department of Agriculture marketing order programs—from peanuts to milk. Carol Foreman (who, if Bob Bergland can be described as "rabidly profarmer," is most certainly "rabidly proconsumer") would have been devastating in such a position. She understands market forces as well as supply and demand. She understands that federal milk regulation is a consumer ripoff designed to protect businessmen-farmers at the expense of the public.

Jimmy Carter was not stupid. Having become a millionaire with the help and assistance of the Department of Agriculture's peanut subsidy program, he would not as president preside over the demise of the source of his personal wealth. Because that is what might have happened if Carol Tucker Foreman had been put in charge of the Department of Agriculture's market orders—a fox guarding the chicken coop. So Carter and Bergland gave Foreman half a job by taking the office of assistant secretary for marketing and consumer services and dividing it into two jobs: an assistant secretary for marketing services and an assistant secretary for consumer services. Foreman was limited to consumer services and was not allowed anywhere near the powerful marketing services position. Carol Foreman, knowledgeable in the ways of official Washington, knew all this before she accepted the position: "I was very carefully excised from that...responsibility."[17] She may have thought she could do more from the inside but, by accepting the job, she gave the Department of Agriculture a patina of consumerism which it neither desired nor deserved.

Some people seemed surprised in 1978 when Jimmy Carter, after giving dairy farmers much more than they wanted in 1977, would take such an otherwise hard nosed position on price supports for other agricultural commodities. His rigid position on wheat and grain price supports resulted in the much publicized, ultimately inconsequential, farmers strike of 1978. The strike essentially involved wheat and grain farmers who wanted either higher market prices or higher price supports and government loans to compensate them for what they believed was a lean year.

Bergland was Carter's point man in reacting to the farmers strike. Meeting with a group of strike leaders in the House Agriculture Committee hearing room, Bergland told them that there was no quick fix or free lunch in farming and that it was not the role of the federal government to guarantee all farmers a profit year after year. Bergland's lecture on the virtues of the free market was in stark contrast to candidate Carter's naive promise in the presidential campaign to guarantee farmers their "cost of production," naive because cost of production is an impossible figure to use in setting policy. What is cost of production for one farmer is not for another, so if cost of production is set high enough, it is bound to benefit many more wealthy farmers who do not need it than poor, inefficient farmers who do.

Bergland's lecture was also in stark contrast to his treatment of the dairy lobby a year earlier, when he and Carter more than guaranteed dairy farmers' cost of production with the milk price support payoff.

The difference between Carter's treatment of wheat and grain farmers, and dairy farmers, therefore, cannot be explained by campaign promises. The promise to the dairy lobby was kept; the promise to wheat and grain farmers was not. The difference also cannot be explained by the relative merits of each decision. Everything Bergland said about there being no "free lunch" in farming and that it was not the "role of the federal government to guarantee all farmers a profit year after year" could have applied equally to dairy farmers. But it was not. The dairy lobby had seen to that. For all but the marginal dairy farmer, the federal government did guarantee dairy farmers a profit year after year.

One reason for this, of course, is political power. Dairy farmers have it. Wheat and grain farmers do not. Tractorcades, indignation, and a Carter campaign promise are simply no match for $1.4 million in campaign contributions.

Another reason, unique to the Carter Administration, is that Carter was a peanut farmer who ran several peanut-connected businesses: a peanut-shelling plant, peanut warehouse, and a seed and fertilizer business. Peanut farmers have little in common with grain farmers, but given their extensive (and expensive) coddling by the federal government, peanut farmers have a lot in common with dairy farmers. As is true with milk, the Department of Agriculture subsidizes peanuts by setting a price support level at which it will buy all surplus peanuts not purchased on the open market. As is true with milk, there are strict import controls for foreign peanuts. The support price in 1976, for example, was approximately $400 a ton—compared to a world price of $250 a ton. The government annually purchases approximately one-third of the peanut crop, some six hundred thousand tons, most of which sit in warehouses and are subsequently given away to needy nations, at a cost to American taxpayers of $155 million a year. Though private purchasers could buy peanuts much cheaper abroad, the Department of Agriculture will not let them. Private purchasers here *must* buy U.S. peanuts. Similarly strict import controls exist for powdered milk and cheese.[18] Even Earl Butz, who did many favors himself for the dairy lobby as Nixon's agriculture secretary, denounced the Department of Agriculture's peanut subsidy program as "a vicious little monopoly."[19]

With such a guarantee, you would expect peanut farming to be a very popular business. And it is, particularly for those who are already peanut farmers. Those who would like to be peanut farmers, however, cannot. Or, at least if they do go into peanut farming without the Department of Agriculture's consent, they will not qualify for a support price and will have to take whatever the market will offer, i.e., approximately $150 a ton less than the Department of Agriculture subsidized price. The Department of Agriculture does this through the device of "allotments," i.e., acreage on which peanuts can legally be planted in order to be eligible for the artificially high price paid by the Department of Agriculture's support program. Peanut allotments, understandably, are in short supply. They total only 1.6 million acres, the same as when the program was created in 1940. All the allotments are taken and, accordingly, are extremely difficult to buy from existing peanut farmers. "If you're in, you're in. If you're out, you can't get in," said Earl Butz.[20]

Jimmy Carter is in. His 228 acre allotment was seven times as large as the national average. John Connally is also in, his 472 acre allotment being fourteen times the national average.[21] It should come as no surprise, therefore, that both Carter and Connally, in their public careers, have shown greater than average understanding and sympathy regarding milk price supports. More than most politicians, they intuitively understand the importance of price supports to a farmer, peanut or dairy. In retrospect, it is obvious that Jimmy Carter, when he used campaign rhetoric about "reduc-[ing] the tremendous increase in the price of farm goods from the farm to the consumer" and "closing the revolving door that exists between the Agriculture Department and the large special interests," had something other than milk or peanuts in mind.[22]

The Bergland-Carter price support payoff is not the only occasion the dairy lobby had during the early Carter years to flex its political muscle. The great ice cream war was another. It all started innocently enough when the Food and Drug Administration (FDA) published revisions in its frozen dessert standards on April 12, 1977. The revised standards required full ingredient labeling for ice cream; provided for the use of any "safe and suitable milk-derived ingredients" in ice cream; and substituted a minimum protein requirement for ice cream, rather than the previous standard of a minimum amount of nonfat milk solids.[23]

Initially, the new standards did not seem controversial. In fact, they represented a new regulatory philosophy at the FDA that food processors should not be denied use of new technologies so long as nutritional standards were not lowered. Hence, the permission to use "all forms" of safe and suitable milk-derived ingredients, and the utilization of a nutritional standard (i.e., a protein minimum) in lieu of the old standard of a minimum level of nonfat milk solids.

Ice cream manufacturers were pleased, because the new standards removed existing artificial restrictions on the amounts of "whey" and "casein" which could be used in ice cream. Whey and casein are cheaper "milk-derived ingredients" than nonfat milk solids, and their use in increased amounts would reduce the retail price of ice cream approximately 2.5 percent. All in all, not a bad deal from a government agency whose prior reputation in Washington for bureaucratic delay and red tape was near legendary: a lessening of overly protective restrictions in the FDA's recipe for ice cream, a cheaper price for ice cream, and strict labeling requirements telling people what is in the ice cream, all with no drop in nutritional value. Everybody should have been happy.

Almost everybody. The dairy lobby was displeased. And in the corridors of power in Washington, it is not nice to displease the dairy lobby. Overnight the dairy lobby began trumpeting a new found concern for the American consumer which it had hitherto effectively concealed.

Dairy lobby flacks began copiously crying public tears about "cheap chemicals" being used to destroy ice cream (whey and casein are chemically derived milk by-products in a process which does not lessen their nutritional value) and "quickbuck artists" (the ice cream manufacturers) trying to destroy an American institution (ice cream). Even Senator William Proxmire (D. Wisc.) got into the act, criticizing the FDA for its new standards.

The FDA was not amused by the dairy lobby counterattack. The FDA is one of the most powerful bureaucracies in Washington with responsibilities for guarding the public health which have made it relatively immune to political attack. It replied in kind to Proxmire. As the *Washington Post* quoted one FDA employee:

I love Proxmire. I think the world of him, but he's being had now. We have not lowered the nutritional or technical profile (of ice cream). We've removed the recipe. If I were a milk producing farmer, I'd be stamping and crying in my beer too. They had a rather guaranteed market.[24]

But the dairy lobby did not just cry in its beer. It moved. In June, 1977, Bob Bergland joined House Agriculture Committee Chairman Tom Foley and Dairy Subcommittee Chairman Charlie Rose in writing to Health, Education and Welfare Secretary Joe Califano (whose department has jurisdiction over the FDA), requesting that the FDA hold hearings on the proposed ice cream standards and delay their implementation. Although no public response was forthcoming from either FDA Commissioner Donald Kennedy or his immediate superior, Califano, Bergland smugly predicted to the press that Califano and Kennedy would hold the requested hearings. Similarly, even though Bergland admitted he had not discussed with Califano and Kennedy whether the FDA would delay the date when the ice cream industry could begin using the new standards, he was confident again that implementation of the new standards would be delayed. And, sure enough, late in June, true to Bergland's predictions, the FDA announced it would not permit the new ice cream standards to go into effect as scheduled. The FDA did not announce, however, which, if any, of the issues raised by the dairy lobby would be considered at the public hearings.[25]

In the meantime, the dairy lobby's retainers on Capitol Hill were hard at work. One congressional dairy loyalist was Charlie Rose, the House Dairy Subcommittee chairman. Rose, in an attempt to publicly (and privately) pressure the FDA into backing off from its new ice cream standards, held his own hearings on the matter. Although the farm population of his district is only 7 percent, Rose had received substantial contributions from the dairy lobby, $9,000 in 1976 alone, with the entire subcommittee receiving almost $90,000. Rose led off his hearings on the FDA's proposed standards by comparing supporters of the new FDA standards to Nazis:

I think what we are concerned about is that America does not go the way of Hitler's Germany in World War II where ersatz became a byword of products that were sold to its people out of economic necessity.[26]

Rose's hearings raised a number of interesting points, most of which Rose was not interested in. The hearings symbolized a clash between those relying on modern food technology to assure commercial success versus the dairy lobby, whose approach to commercial adversity invariably involved some form of government protection at the expense of consumers. Witnesses such as M. E. Gregory, a food science professor from North Carolina State University, and Peter Hutt, former general counsel of the FDA, emphasized that the dairy lobby's reaction to the new ice cream standards was self-defeating. Hutt echoed Gregory in explaining that he had seen the dairy industry lose the butter market to margarine, the aerated whipped cream market to whipped toppings and the cream market to coffee whiteners.[27] A major factor in each of these examples was the dairy industry's attempt to rely on legal protection rather than take advantage of modern food technology to market new products that would be more appealing to consumers. Hutt suggested that if the dairy industry did not learn from its past mistakes, it could also lose the market for ice cream.

FDA Commissioner Donald Kennedy also appeared at the Rose hearings and denied dairy lobby charges that the FDA was allowing a deterioration in the quality of ice cream:

> What happens, as happens in any technology, is that people find out how to do something a little better within the old limits, and then people exercise their preferences in the market place. There is plenty of ice cream sold that is absolutely delicious. Just costs a lot to make just as it costs you a lot to make at home if you choose to be a rebel against the quality of ice cream that is in the supermarket.[28]

More important, Commissioner Kennedy did not act subserviently at Rose's hearings. He told the subcommittee that the FDA had no intention of granting a hearing on its new frozen dessert standards unless the dairy lobby could show that the standards would result in a less nutritious product or that prices to consumers would be adversely affected. Kennedy was unconcerned with the adverse effect, if any, on dairy farmers, and suggested that the FDA's obligations were to the food consumer, not the dairy lobby.

Rose was unhappy with Kennedy's approach. Everyone on Capitol Hill knew that the dairy lobby was more important than consumers. Joe Califano knew. Charlie Rose had flown back from Paris with Califano after an international dairy conference and told him so. But the FDA still did not understand. Rose decided that this lesson ought to be taught more forcefully. After the hearings, Rose announced that he would focus on legislative efforts to reverse *all* food standards issued by the FDA, using its "safe and suitable" and "nutritional equivalence" approaches. Rose talked pompously about "reversing" the damage already done by FDA standards which "foist inferior foods off on an unknowing public."[29] Senators Humphrey and Talmadge promptly lined up behind Rose's endeavor (although no senators from committees with jurisdiction over the FDA joined them).

In the meantime, the dairy lobby was having a difficult time on the merits. Pressed by the FDA to prove its claim that the new ice cream standards would produce inferior ice cream, the dairy lobby admitted it could not do so. It is up to the FDA, said the dairy lobby, to prove that the new standards will not produce inferior ice cream.[30]

At the same time, the real motives of the dairy lobby began to surface publicly. The purpose in opposing the FDA ice cream standards was to protect the price support program and the price of raw milk generally, i.e., a decreased use of milk powder in ice cream would result in an increased milk surplus, which would have to be purchased by the government at an additional cost in excess of $200 million a year. Because this came on top of the Carter-Bergland price support payoff and the Omnibus Farm Bill, the dairy lobby feared that paying another $200 million a year to subsidize dairy farmers might be too much and that such a shock might make Congress rethink the funding of the entire milk price support program.

By October of 1977, the FDA had received more than six hundred comments requesting a hearing on its ice cream standards. The National Milk Producers Federation had filed a declaratory judgment action in federal court to require the FDA to consider the narrow private interests of the dairy lobby (by taking into account the economic impact of the proposed standards on dairy farmers) rather than focus solely on the FDA's traditional concerns with the general welfare of American consumers. While Kennedy bravely attacked Congress for its response to the new FDA standards as a "spasm of dairy protectionism," he publicly admitted that the FDA was going to reconsider its new ice cream standards.[31]

On December 16, 1977, the FDA withdrew all of its new ice cream standards except the one requiring mandatory ingredient labeling. The FDA explained that its new ice cream standards might "result in a reduction in nutrient levels in some frozen desserts" and that the rationale for this conclusion would be released later in the *Federal Register*.[32] It never was. The rationale subsequently published by the FDA in the February 3, 1978, *Federal Register* was no more specific than its original decision—an implicit admission that Kennedy and his boss Joe Califano had made a purely political decision. As the *Dairy Industry Newsletter* reported:

> In three short and nonspecific sentences, the agency explains that after evaluating "data received in response to the . . . request for information" and conducting its own studies it has concluded that the revised standards could result in a reduction in nutrient levels in some frozen deserts.[33]

That was it. Nothing more. There was no justification. Ice cream manufacturers plaintively complained that it was "a politically expedient sidestepping of the issues."[34] But calling it expedience is putting too kind a face on it. It was an exercise in raw political power against one of the most powerful entrenched bureaucracies in Washington. The bureaucracy lost. The dairy lobby won. After only three years and $1.4 million in contributions, the dairy lobby had successfully risen from the ashes of Watergate. Its victory in the great ice cream war demonstrated that rise for all to see.

11

Real Enemies, Potential Friends: The Prospects for Reform

The dairy lobby's victory in the ice cream war was an important one. Political power in Washington is ephemeral unless it is constantly nurtured and reinforced. A special interest group like the dairy lobby cannot afford even the appearance of weakness on issues important to it. If power is to be maintained, enemies must be impressed by that power and friends reinforced in their support.

The most impressive aspect of the dairy lobby's performance during the 1970s was that it survived its seamy role in Watergate with its power intact. A lesser lobby would have folded its tent in embarrassment and headed for home. The dairy lobby did not. Watergate was its baptism of fire, and it learned its lesson well.

In retrospect, the dairy lobby's strategy was a high risk one, but well conceived. A low profile was understandably called for in 1974: campaign contributions were drastically cut back, and the bulk of these were given either after the election or after the last reporting date prior to the election. The surprise was the lobby's spurning of a gradualist approach in 1976. Since the 1974 Federal Election Campaign Act had severely limited the dairy lobby's ability to make massive contributions to presidential candidates, it had that much more money available for congressional candidates. The dairy lobby simply decided to spend almost all of it on Congress—$1.4 million—and literally buy its way back to respectability.

This may seem paradoxical to some but not to those familiar

with life in Washington. Les Whitten, an investigative reporter and associate of Washington columnist Jack Anderson, offered an explanation in the opening scene of his 1976 political novel *Conflict of Interest*:

> The reception was given by the milk producers, once the center of such scandal in our city that you would think a speaker of the house and his wife would never again be their guests. But in our compromised Capitol, all things can be forgiven except for certain overt types of disloyalty and persistent rudeness.[1]

The political victories of the dairy lobby in 1977—Carter's price support payoff, the Omnibus Farm Bill, and the ice cream war—showed indeed that all things can be forgiven in Washington. All it takes is lots of money—sums made possible every time someone buys a container of milk—and a willingness to abide by the new rules on how the money is to be distributed.

The lesson of the dairy lobby's ability to maintain power in Washington, even during the darkest days of Watergate, has not been lost on the dairy lobby's potential enemies. Edward Kennedy is a good example. Fresh from efforts at initiating deregulation of the airlines (whose political power he had correctly calculated as being less than overwhelming), Kennedy had seemingly set his sights on the dairy lobby as his next victim. It was early in the morning on a cold December day in 1975 and he was the featured speaker at a Washington conference on milk marketing reform. Kennedy's face was florid; he drank two glasses of water waiting for his introduction; and he read from a prepared text. But it was a good speech, perceptive in its understanding of why the price of milk is excessively high, and the people from the dairy lobby who heard it squirmed with discomfort in their chairs:

> As all of us recognize, milk and dairy products play a central role in our daily diet. At the same time, they also represent a significant portion of our nation's grocery bill. This is no small consideration these days.
>
> For today the overriding issue across our land is the country's economic plight and the role of the federal government in dealing with the current crisis. The inability of government to keep the economy on an even keel—with stable prices and full employment and reasonable interest rates—has resulted in not only public disillusionment, but massive cynicism and outright distrust of government. This is destroying our nation's basic faith and con-

fidence today, and the people are right when they blame government. . . .

The problems of our economy have occurred not as an outgrowth of laissez-faire, unbridled competition. They have occurred under the guidance of federal agencies, under the umbrella of federal regulations. . . .

Both Ralph Nader and Ronald Reagan seem to agree that there is too much government interference in the marketplace these days. That should give all of us food for thought. . . .

The task of regulatory reform requires the fullest cooperative efforts of both the Congress and the president. That task is long overdue. It should not get caught up in the stalemate of election-year politicking.

Over the past year, the subcommittee on Administrative Practice and Procedure, which I chair, has been looking at the Civil Aeronautics Board. We have been asking what that regulatory agency is doing, how it is working, and how it might better respond to the needs of the industry and the public. A central question running throughout our hearings was why haven't American consumers been provided the choice of low-cost air service as an alternative to the high-priced service now available.

In addressing this question, we tried to separate fact from myth, and that was the biggest challenge of all.

We were told by industry spokesmen that airline regulation was first imposed by Congress to overcome the destructive, cutthroat competition which had prevented development of a unified, dependable network of air service previously.

Yet our own review of both legislative history of the Federal Aviation Act and economic history of the industry revealed that Congress passed that law in response to growing industry concentration and financial losses which were caused not by competition but by the way in which the federal mail subsidy program had been administered up to that time.

We were told that in a competitive environment airlines would be unable to generate excess profits on some major routes to support unprofitable service to small communities, and thus economic development of many parts of our country would be impaired.

Yet when I asked each airline to provide a list of the communities that would lose service in a competitive environment, only the most feeble response was forthcoming.

We were told that intrastate carriers in Texas and California offer service at down to half the cost of similar flights in the East because of such factors as weather, passenger volume, delays, labor costs, and others.

Yet in our hearings we compared these factors in great detail and determined that there was really only one factor which explained the difference between the approximately $20 Los An-

geles-San Francisco fare and the $45 Washington-Boston fare:
federal regulation of interstate carriers.

In short, once the myths were dispelled, we found that federal
economic regulation of the airlines is responsible for stifling price
competition, maintaining most air fares at excessively high levels,
and costing consumers anywhere from $1 to $3.5 billion every
year. We also learned that neither small communities nor airline
companies themselves benefit in any real economic sense from
federal regulation. Instead, the lion's share of these excessive costs
go to pay for hundreds of half-empty planes flying around the
country.

From my initial review of the milk marketing system, some of
the same disturbing symptoms are appearing.

A number of studies have begun to unravel the maze of federal
regulation, only to reach the conclusion that its costs—to both the
dairy farmer and the consumer—are far greater than its benefits.

There is something wrong with a system where the government
spends millions of dollars on supports to keep prices up, while
those same high prices force demand down to the point that
thousands of producers go out of business during the same period.

There is something wrong with a system that results, according
to some agricultural economists and government agencies, in
$200 to $400 million in consumer overcharges for milk each year.

There is something wrong with a system that first sets regulated
prices artificially high, and then allows some monopolistic cooperatives to extract premiums over and above those prices.

There is something wrong with a system that was intended to
provide indirect subsidies to hard-pressed dairy farmers and instead transfers money from milk consumers to large agricultural
corporations and rural mortgage lenders. . . .

[M]ilk marketing regulation has . . . gone astray.

Those who defend the present system, therefore, must identify
clearly and accurately the benefits of the system. It may be, as I
found in the field of airline regulation, that only a small fraction
of those consumer dollars actually end up in the hands of those
intended to be benefited by regulation. . . .

With milk products as with air travel, the watchful eyes of price
competition and consumer demand may well be better regulators
than a handful of civil servants in Washington.[2]

The dairy lobby naturally did not agree. It had long before
learned that it was far easier to manipulate that "handful of civil
servants in Washington" than the American consumers. Kennedy's
speech had caught dairy lobby officials by surprise but they moved

quickly to respond. Smoothly. Quietly. No strident speeches attacking Kennedy's heartlessness. Instead they went directly to Kennedy's staff, armed with their own self-serving arguments to prove that their industry really *was* different and that dairy farmers could never survive without government regulation. They also brought with them a list of the dairy lobby's friends in the Senate—powerful figures like James Eastland, chairman of the Judiciary Committee, and Herman Talmadge, chairman of the Agriculture Committee, respected liberals like Hubert Humphrey, George McGovern, William Proxmire, Gaylord Nelson, and even Kennedy's old Harvard roommate, John Culver of Iowa.

Kennedy and his staff got the message. Within a short time, Kennedy began to hedge. While still claiming that an investigation of milk marketing was under consideration, his staff told journalists that milk marketing was more complex and the answers less clear than those in airline regulation. His staff also said they thought the Department of Agriculture bureaucrats—for decades unrivaled in Washington for sheer sloth and bloat—were doing a "good job" compared to those at the Civil Aeronautics Board. Finally, his staff disclosed that Kennedy was "more in sympathy" with dairy farmers than with airline executives.

It was a tacit admission that Kennedy and his staff had simply underestimated the power and influence of the dairy lobby on Capitol Hill. They obviously had believed that the dairy lobby would be as easy and vulnerable a target as the airlines. When they learned otherwise, Kennedy quickly backed off and never gave another speech on milk marketing reform.

It would be an exaggeration to suggest that the dairy lobby was actually able to intimidate Kennedy into retreating from milk reform. Nevertheless, Kennedy is a keen student of political reality and well aware of how thankless a task deregulation of any industry can be. "You're making real enemies today," he has said, "for potential friends tomorrow."[3]

Indeed, Kennedy's reluctance to take on the dairy lobby contrasted with his position on trucking deregulation affords a direct comparison of the dairy lobby's relative power today in Washington. In 1979, from his new position as chairman of the Senate Judiciary Committee, Kennedy turned his attention to trucking deregulation, an issue strongly opposed over the years by the powerful highway and trucking lobby, which includes the country's largest labor union, the International Brotherhood of Teamsters. That Kennedy would, in his deregulatory zeal, risk offending the highway lobby and the Teamsters rather than the dairy lobby is eloquent testimony to its continuing power today.

John Seiberling, a congressman from Akron, Ohio, is also well aware of the dairy lobby's continuing political power. A respected antitrust lawyer before his election to Congress, Seiberling is a liberal Democrat who serves on the House Judiciary Committee. More than most of his colleagues, he is familiar with the major role played by the dairy lobby in the Judiciary Committee's impeachment investigation of Richard Nixon.

While the House Judiciary Committee has no jurisdiction over milk price support and milk marketing legislation, it does have jurisdiction over antitrust matters. As a consequence, Seiberling introduced in 1977 a moderate antitrust reform bill which, if enacted, would have limited what agricultural co-ops believe to be the scope of their exemption from antitrust laws. It would have made all co-ops subject to the same antimerger laws which govern business corporations, and it would have outlawed the blatant price-fixing engaged in by many co-ops and otherwise approved by some courts. But the bill never got off the ground, and Seiberling has been unable to persuade House Judiciary Chairman Peter Rodino to even schedule hearings on it. This is the same Peter Rodino who received $4,100 from the dairy lobby in 1974, even though his total campaign expenses that year were only $6,286, and his hapless GOP opponent received only 15 percent of the vote.

Other reform measures have met a similar fate. The most comprehensive were those proposed by the Ford Justice Department in its massive six-hundred-page analysis of federal milk marketing orders and price supports. The lawyers and economists who drafted the report declared that milk price supports were "economically invalid and inefficient." They recommended as a reform measure the gradual deregulation of all federal pricing of milk. The Department of Agriculture would still verify weights, and tests and keep records but would no longer price milk.

Computer simulations run by the economists showed that deregulation would result in consumers saving at least $200 million a year.

Another remedy proposed by the Justice Department economists was to lower the Department of Agriculture's artificial difference in price between Grade A milk and Grade B milk. This would have the immediate effect of lowering the chronic Grade A surplus of milk in this country while, at the same time, assuring an adequate supply of Grade A milk for all markets. The Department of Agriculture could accomplish this virtually overnight by ceasing to assume that its legislative mandate of providing an adequate supply of milk implicitly means an adequate supply of *locally* produced milk. Given today's transport technology, milk can easily be shipped over long distances—from Minnesota and Wisconsin to Florida for example. Accordingly, areas of the country which would not produce a milk surplus without the artificial stimulus and protection of federal regulation would acquire their milk from areas like Minnesota and Wisconsin, which do have a natural surplus.

When Jimmy Carter took over, he moved quickly to kill the Ford reforms. Carter's attorney general, Griffin Bell, fired the two antitrust division lawyers who had been most publicly critical of dairy cooperatives and their exemption from the antitrust laws. "There's a lot of merit," drawled Judge Bell, in response to later questions about the sacking of the two, "in change for change's sake."[4] Agriculture Secretary Bob Bergland openly attacked the Justice Department report for its "factual, legal and interpretative errors."[5] When Bell failed to allow the Justice Department to respond to Bergland's public criticism, the dairy lobby knew that milk reform was dead and that another Carter campaign commitment had been met.

In December 1977, Carter appointed a presidential commission to review antitrust laws and procedures. Its mandate included a review of antitrust exemptions and immunities for agricultural cooperatives. The twenty-two-member commission consisted of five senators, five representatives, three heads of federal agencies, one federal district court judge, one state attorney general, and seven private sector attorneys with academic and antitrust backgrounds. While its members included Edward Kennedy and John Seiberling, the very nature of such commissions is to afford the illusion but not the reality of reform. And in case that was not clear to the commission when he created it, Carter made sure they got the message that summer when he made a highly publicized farm speech in Columbia, Missouri, defending the antitrust exception for agricultural cooperatives. "I will always protect the Capper-Volstead Act," the president declared.[6]

The commission got the message. It released its report early in 1979, and its recommendations on agriculture were predictably mild. First, it said mergers and price fixing among cooperatives should continue to be permitted so long as "substantial" lessening of competition did not result. Second, it recommended that the Capper-Volstead prohibition against "undue price enhancement" of agricultural products (a prohibition never enforced by the Department of Agriculture in over forty years) not be administered by the same part of the Department of Agriculture which is responsible for promoting the well-being of cooperatives. On the only issue before it which had any real meaning for consumers—the exemption from the antitrust laws of federal marketing orders for milk and other agricultural commodities—the commission did nothing: "The commission is not able to make a definitive recommendation concerning the current exemption for agricultural marketing orders."[7]

The dairy lobby had understandably made sure the commission was aware of its views. In an article in the October, 1978 issue of *American Agriculturalist,* the president of the National Council of Farmer Cooperatives claimed that "Farmers are facing one of the greatest threats to their cooperatives of any period in recent history. . . . We'll have to work together—cooperative members, employees, and supporters alike—to defend our role in the American free enterprise system. The challenge is up to us. If we meet it well, farmers, consumers, and our country will be the better for it."[8]

The challenge was met, and consumers were given scant consideration by the commission. Dairy lobby representatives who appeared before the commission boldly stated that "the milk marketing system . . . by its very nature . . . is designed to transfer income from consumers to [dairy farmers]." Essentially, the dairy lobby told the commission to mind its own business and stay out of "political" areas where the dairy lobby is free to gouge billions of dollars each year from taxpayers and consumers. The commission meekly agreed: "The commission has not attempted to pass judgment on the *pure transfer aspects* of the program; whether consumers should continue to subsidize producers is a *political question* for Congress" (emphasis added).[9]

Meanwhile, back in Congress, no one was listening to the commission anyway. If Congress was listening to anyone, it was to Gaylord Nelson, the junior senator from Wisconsin. When asked how Congress would look at any legislation dealing with changes in the nature of dairy cooperatives or the federal milk marketing system, Nelson's answer was that "most members of Congress would look at the question from a position of profound ignorance." Nelson candidly admitted he would oppose "any proposal that would weaken the market position of the dairy farmers." Further, Nelson did not believe that the dairy lobby was getting enough:

> It is the view of many consumers (perhaps a substantial majority) that retail prices can be appreciably reduced by further squeezing the dairy farmer. They believe that the dairy support program is an indefensible expenditure of taxpayers' money to subsidize the farmer. This, despite the fact that the dairy farmer and his family work longer hours each day, more days a week, for lower average income and less return on investment than any other class of businessmen in our economy. Not infrequently, consumers, city folks, members of Congress, ask me why they should continue to subsidize the dairy farmer. My response puzzles them, convinces no one, but makes me feel better. I point out that they are asking the wrong question. The right question is: How much longer can the dairy farmer be expected to continue to subsidize the consumer? When one efficiently produces a good product for a low return on one's labor and frequently little or no return for investment or management, it is the producer who is paying the subsidy, not the consumer.[10]

Nelson knows better than this. When consumers are willing to pay a certain amount for dairy products, freely, on the open market, without someone's holding a gun at their heads and lifting their wallets, and dairy farmers are receiving (as they traditionally have) well over 50 percent of every dollar spent on dairy products, economists generally concede this to be a "fair" price. If it is not fair (i.e., if the vast majority of dairy farmers do not make a good or high return on their own labor and little or no return for their investment), economists respond that either (a) there are too many dairy farmers or (b) the dairy farmers we have are producing too much raw milk. In a free market, overproduction stops when prices fall. What happens under the present system is that when prices start to fall, government intervenes and props up prices by taxing consumers and using the proceeds to buy up the resulting surplus milk. Such a system fits the classical economic definition of a government subsidy, a government subsidy *paid by the consumer,* not (as Senator Nelson, in his colorful imagination, suggests) a subsidy paid by the producer.

The senior senator from Wisconsin, William Proxmire, frequently engages in similar economic sophistry on behalf of the dairy lobby:

> . . . The No. 1 victim of economic injustice in this country today is the farmer and especially the dairy farmer. He does everything a person is supposed to do to justify a reasonable income in a free economic society: He takes a big risk, not only on prices but on weather; he makes a large investment of capital; he works tremendously long hours. . . . And his reward for all this is an income that is an insult.[11]

Financial journalist James Grant found a valuable lesson in all this:

> Broad application of Proxmire's idea would provide welcome news to Xerox Corp., which invested $1 billion to compete with IBM in computers, only to fail. . . . By dint of hard work and a willingness to take risks, the senator suggests, the dairy farmer has established a moral claim on the U.S. treasury.[12]

This is the same Senator Proxmire who developed the "Golden Fleece" award, which he regularly bestows with appropriate publicity on bureaucrats who devise novel little schemes to waste taxpayers' money. Billion dollar rip-offs like federal milk price supports are apparently not nearly as novel or newsworthy. Proxmire's attitude, however, is symbolic of the prospects for political reform of milk marketing in this country. Consumer groups simply do not have $1.4 million to give away every two years to offset the influence and access which the dairy lobby continues to buy.

New evidence of this was furnished as recently as November, 1979, when the House passed an amendment to the Federal Trade Commission Improvements Act of 1979 that prohibited the Federal Trade Commission (FTC) from using funds to study, investigate or prosecute any agricultural cooperative. The amendment was offered by Representative Mark Andrews, a Republican from North Dakota who had frequently received contributions from all three of the dairy lobby's political action committees. The amendment had considerable bipartisan support, however, particularly from the California delegation, because it would have the effect of killing the FTC's current antitrust case against Sunkist Growers, Inc., a large California-based cooperative charged with illegally monopolizing the western citrus fruit industry. It would also kill pending formal investigations against Dairymen, Inc. and the Ocean Spray cranberry-growing cooperative. Hearings were never held on the Andrews amendment, and no testimony was ever taken for or against it. It was a pure power play by the dairy lobby and their agricultural cooperative allies like Sunkist. It was offered at the last minute and brought to the House floor under a modified closed rule (a highly unusual procedure for an amendment on which hearings had not been held) so that no alteration or clarification to the Andrews amendment could be offered on the floor of the House.

The Andrews amendment passed by a two-to-one margin, and the dairy lobby had taken a giant step toward assuring that there would be no effective reform for milk marketing in the 1980s. Unlike the Department of Justice and the Department of Agriculture, the FTC had been relatively immune from the political pressure of the dairy lobby. Charged with protecting consumer interests, the FTC had not only investigated and prosecuted agricultural cooperatives, it had also investigated the anticompetitive, anticonsumer nature of marketing orders in milk, citrus fruit and other commodities. The Andrews amendment would effectively stifle the only independent voice in official Washington speaking out for pro-consumer reform on food policy issues.

The reforms needed in milk have long been evident to anyone who objectively studied the problem: (1) eliminate milk price supports and the resulting huge excessive milk surplus built up at taxpayer expense; and (2) abolish federal milk marketing orders, which extract unreasonably high milk prices from consumers.

These reforms would work. Econometric models prove it. More milk would be sold at lower prices. Efficient dairy farmers like those in Minnesota and Wisconsin would receive more for their milk than they do today. Marginal dairy farmers would continue to leave dairy farming, just as they do today. The swollen Department of Agriculture bureaucracy could be reduced. Taxes could be cut.

But there is no political constituency in Washington today for such reform. And without such support, the dairy lobby will continue its domination of Congress. Gaylord Nelson may be correct when he says that most congressmen view the issue of milk marketing reform with "profound ignorance" but it is more accurate to observe that on this issue there are few politicians in Washington or elsewhere who are willing, in Edward Kennedy's words, to make "real enemies today for potential friends tomorrow." A dairy lobby which continues to spend $1.4 million on every election will deter all but the most intrepid congressmen from supporting milk reform.

The abuses inherent in a government influenced by special interest groups is not a new phenomenon. Politicians who rob Peter to pay Paul can always depend upon the support of Paul.[13] Perhaps if individual congressmen were prohibited from voting on legislation conferring economic benefits on members of special interest groups which contributed to their election campaign, the public would no longer be regularly fleeced every time Congress is in session. How many millions would the dairy lobby spend on election campaigns if it knew the recipients were prohibited from voting to increase milk price supports? As it is, narrow economic interests like the dairy lobby depend upon a return on their investment when they spend $1.4 million every election and no one should be surprised when those grateful politicians subsequently gouge billions of dollars from the pockets of consumers and taxpayers and transfer it to their dairy farmer benefactors. Lincoln Steffens had it right: "Power is what men seek, and any group that gets it will abuse it. It is the same old story."[14] It is a story learned and practiced well by the dairy lobby.

Chapter Notes

INTRODUCTION

1. "Between the Lines," *Monthly Detroit* (April, 1978) p. 6.

NOTES FOR CHAPTER ONE

1. "Milk: Why Is the Price So High?" *Consumer Reports* (January 1974), p. 69.
2. Ronald J. Knutson, *Cooperative Bargaining Developments in the Dairy Industry 1960-1970*, Farmer Cooperative Research Report No. 19 (U.S. Government Printing Office, 1971), p. 4.
3. *Statement of Information*, Hearings before the Committee of the Judiciary, House of Representatives, 93rd Cong., 2d Sess. [1974], Book VI, Part II, pp. 569, 570.
4. Frank Wright, personal letter.
5. *Statement of Information*, pp. 631-35.

NOTES FOR CHAPTER TWO

1. Mark Kramer, "Making Milk," *Atlantic Monthly,* November 1977, pp. 80-81.
2. Ibid., p. 87.
3. F. A. Hayek, "The Constitution of Liberty" (Chicago: University of Chicago Press, 1960), pp. 359-60.
4. Donald L. Kemmerer and C. Clyde Jones, *American Economic History,* (New York: McGraw-Hill, 1959), p. 419.
5. Ibid, p. 432.
6. Ibid, p. 433.
7. Ibid., p. 434.
8. Ibid.
9. John D. Hicks, *The Populist Revolt* (Lincoln, Nebraska: University of Nebraska Press, 1961), p. 54.
10. Howard S. Russell, *A Long, Deep Furrow, Three Centuries of Farming in New England* (Hanover, New Hampshire: University Press of New England, 1976), p. 440.
11. X. A. Willard, "Beginning of a Cheese Cooperative," *Readings*

in *History of American Agriculture,* Wayne Rasmussen, ed. (Urbana, Illinois: University of Illinois Press, 1960), p. 93.

12. Russell, p. 474.
13. Roger W. Fones, Janet C. Hall, and Robert T. Masson, *U.S. Department of Justice Study of Milk Marketing* (Washington: U.S. Department of Justice, 1976), p. 32.
14. *Ford, et al. v. Chicago Milk Shippers Association* 39 N.E. Rptr. 651 (Supreme Court of Illinois, 1895).
15. George Soule, *Economic Forces in American History* (New York: William Sloane Associates, 1952), pp. 372, 373.
16. Ibid., p. 375.
17. Edward C. Kirkland, C., *A History of American Economic Life* (New York: Appleon-Century-Crofts, 1951), p. 590; see also Soule, 377.
18. Grant McConnell, *The Decline of Agrarian Democracy* (Berkeley and Los Angeles: University of California Press, 1953), p. 363; George N. Peek and Hugh S. Johnson, "Peek and Johnson Advocate a Two Price Plan" in Rasmussen, ed., *Readings in the History of American Agriculture,* (Urbana: Univ. of Illinois Press, 1960), pp. 227, 228.
19. Arthur M. Schlesinger, Jr., *The Crisis of the Old Order* (Boston: Houghton Mifflin, 1957), p. 107.
20. Ibid.
21. Ibid., 108.
22. Russell, pp. 493, 494.
23. 61 Cong. Rec. 1033 (1921).
24. 62 Cong. Rec. 2123 (1922).
25. Reuben A. Kessell, "Economic Effects of Federal Regulation of Milk Markets," *Journal of Law and Economics* (April 1967), p. 52.
26. Fones, Hall, and Masson, p. 77.
27. Schlesinger, *Crisis of the Old Order,* pp. 266, 267; Kramer, Dale, *The Wild Jackasses,* (New York: Hastings House, 1956), pp. 226-228.
28. Kramer, Dale, pp. 229, 230.
29. Ibid., p. 230.
30. Hearings on H.R. 3835, Agriculture Emergency Act to Increase Farm Purchasing Power, Before the Senate Agriculture and Forestry Committee, 73rd Cong. 1st. p. 326 (1933).
31. Fones, Hall, and Masson, pp. 95-98.
32. Schlesinger, *The Coming of the New Deal,* p. 76.
33. Senate Report No. 1011, 74th Cong. 1st. Sess. (1935), p. 3.
34. Edwin G. Nourse, *Marketing Agreements Under AAA* (Washington, D.C.:The Brookings Institution, 1935), pp. 316, 317 (emphasis added).

NOTES FOR CHAPTER THREE

1. Alden C. Manchester, *Pricing Milk and Dairy Products: Principles, Practices and Problems* (Washington: U.S. Government Printing Office, 1971), p. 2.
2. Reuben A. Kessell, "Economic Effects of Federal Regulation of Milk Markets," *Journal of Law and Economics* (April 1967), p. 60.
3. Rowland W. Bartlett, "Bringing Federal Order Class I Pricing Up to Date and in Line with Antitrust Regulations," *Illinois Agricultural Economics*, 14, no. 1 (January 1974), p. 2.
4. Milton Friedman and Robert V. Roosa, *The Balance of Payments: Free Vs. Fixed Exchange Rates* (Washington, D.C.: American Enterprise Institute, 1967), p. 1.
5. Tanya Roberts, *Review of Recent Studies* (Washington, D.C.: Public Interest Economic Center, 1975), pp. 25-32.
6. David R. Fronk, *Farmer Size and Regional Distribution of the Benefit Under Federal Milk Market Regulation*, Bureau of Economics Staff Report to the Federal Trade Commission (Washington, D.C., May 1978), p. 21.
7. Roberts, pp. 41-43.
8. Ibid., pp. 36-40.
9. Ibid., p. 44.
10. Ibid.
11. Ronald D. Knutson, *Cooperative Bargaining Developments in the Dairy Industry 1960-70*, Farmer Cooperative Research Report No. 19 (Washington, D.C.: U.S. Government Printing Office, 1971), p. 4.
12. Randall E. Torgerson, *Producer Power at the Bargaining Table, A Case Study of the Legislative Life of S. 109*, (Columbia, Missouri: University of Missouri Press, 1970), p. 58.
13. Robert E. Jacobson, "Dairy Co-Op Members and Buyer Discrimination," *Modern Milk Marketer* (June 1975), p. 4.
14. Knutson, p. 10.
15. Ibid., pp. 10, 12; Fones, Hall, and Masson, pp. 166-79.
16. Ibid.
17. Ibid.
18. Ibid.; *Milk Reporter* (October, 1977).
19. Robert E. Jacobson, "Legal and Economic Implications of the New Dairy Cooperative Structure," *Proceedings of Ohio Dairy Seminar*, Cooperative Extension Service, Ohio State University (Columbus, Ohio: April, 1971), p. 30.
20. Ibid., p. 32.
21. Knutson, pp. 15-22.
22. Ibid.
23. Ibid.

24. Edith Hall Parker, "Monopoly in the Milk Producing Industry: the 'Super Cooperatives,'" *Antitrust Law and Economics Review* (Summer 1970), p. 118.
25. Manchester, p. 44.
26. Phillip Eisenstadt, Robert T. Masson, and David Roddy, *An Economic Analysis of the Associated Milk Producers Inc. Monopoly* (Washington, D.C.: U.S. Department of Justice, 1974), p. 125.
27. Brooks Jackson, "Milking the White House," *New Republic,* January 26, 1974, p. 15.
28. Ibid.
29. Fones, Hall, and Masson, pp. 398-99.
30. U.S., Congress, Senate, Select Committee on Presidential Campaign Activities, Hearings, Presidential Campaign Activities of 1972, 93rd Cong., 1st and 2nd Sess., 1974, p. 6907 (hereinafter referred to as *Senate Watergate Hearings*); U.S., Congress, Senate, Select Committee on Presidential Campaign Activities, Hearings, President Campaign Activities of 1972, 93rd Cong., 1st and 2nd Sess, *Final Report,* Report No. 93-981, 93rd Cong. 2nd Sess., 1974, pp. 583-585 (hereinafter referred to as *Senate Watergate Report.*
31. *Senate Watergate Report,* pp. 585-86.
32. Ibid., pp. 587-88; *Senate Watergate Hearings,* p. 5862.
33. *Senate Watergate Hearings,* pp. 6502-3, 6755-56.
34. Ibid., pp. 6379-81, 6509-10, 6757.

NOTES FOR CHAPTER FOUR

1. *Report of Wright, Lindsey and Jennings to Board of Directors of Associated Milk Producers.* (Little Rock, Arkansas: 1974), p. 8.
2. Ibid., pp. 127-32.
3. *New York Times,* March 27, 1974, p. 27.
4. *Senate Watergate Report,* pp. 734-35.
5. *New York Times,* March 27, 1974, p. 27.
6. *Hubert H. Humphrey: The Education of a Public Man* (Garder City New York: Doubleday, 1976), pp. 179-80.
7. *Report of Wright, Lindsey and Jennings,* pp. 40-41.
8. Ibid., p. 110.
9. "Illegal Donations by Dairy Combine Tied to Both Nixon, Humphrey Funds," *Miami Herald* (March 15, 1974), p. 34A.
10. Watergate Special Prosecution Force Memorandum, re. Prosecution of Jack Chestnut (December 10, 1974), p. 2.
11. Ibid.
12. *New York Times* (March 28, 1974), p. 27.

13. *Senate Watergate Report*, pp. 869-72.
14. Ibid., p. 871.
15. Ibid.
16. Ibid.
17. Ibid., p. 891-92.
18. Irwin Ross, "Most Powerful Man in Congress," *Reader's Digest* (January 1971), pp. 104-5.
19. Michael Barone, Grant Ujifusa, and Douglas Matthews, *Almanac of American Politics 1978* (New York: E. P. Dutton, 1977), p. 43.
20. John F. Manley, *Politics of Finance* (Boston: Little, Brown, 1970), p. 125.
21. Ibid., p. 112.
22. Hugh Sidey, "The Republic of Wilbur Mills," *Life* (February 19, 1971), p. 4.
23. "Nixon v. Mills: Showdown on Trade Policy," *Time* (March 22, 1971), pp. 70-71.
24. *Senate Watergate Report*, p. 903.
25. Marshall Frady, "The Wooing of Wilbur Mills," *Life* (July 16, 1971), p. 52B.
26. "Wilbur Mills: Aiming for the White House," *U.S. News and World Report* (June 7, 1971), p. 39.
27. Ibid.
28. *Senate Watergate Report*, pp. 906–7.
29. Ibid.
30. Ibid.
31. Ibid.
32. Ibid.
33. Ibid., p. 903.
34. Ibid., p. 905.
35. Ibid., p. 921.
36. Ibid.
37. Ibid., p. 922.
38. Ibid.
39. Ibid.
40. Ibid., p. 908.
41. Ibid.
42. Ibid.
43. Ibid.; *Report of Wright* et al. pp. 115–17.
44. Ibid., p. 118.
45. *Senate Watergate Report*, p. 913.
46. Ibid., p. 914; *Report of Wright* et al. pp. 117-18.
47. *Senate Watergate Report*, p. 915.
48. Ibid., p. 909.
49. Ibid.

50. Ibid., p. 910.
51. Ibid., pp. 922-23.
52. *Report of Wright*, et al. p. 119.
53. *Senate Watergate Report*, p. 916.
54. Ibid.
55. Ibid., pp. 916-17.
56. Ibid., p. 917.
57. Ibid.
58. Ibid., pp. 917-18.
59. Ibid., p. 918.
60. Ibid.
61. Ibid., p. 928.
62. Ibid., p. 929.
63. Ibid., pp. 903-4.
64. Ibid., p. 907.
65. *Senate Watergate Report*, p. 904.
66. *Senate Watergate Hearings*, p. 1201.
67. *Senate Watergate Report*, p. 920.
68. "A Firecracker Explodes," *Newsweek* (October 21, 1974), p. 42.
69. Ibid.
70. Ibid.
71. Ibid.
72. Ibid.
73. "Wilbur In Nightown," *Newsweek* (December 16, 1974), p. 21.
74. "The Fall of Chairman Wilbur Mills," *Time* (December 16. 1974), pp. 22-23.

NOTES FOR CHAPTER FIVE

1. *Senate Watergate Hearings*, pp. 6385, 6519.
2. Ibid., p. 6385.
3. Ibid., pp. 7188-89.
4. Ibid., p. 7189.
5. Ibid.
6. Ibid., pp. 7190-91.
7. Ibid., pp. 6386, 7191.
8. Ibid., p. 7577.
9. Ibid., p. 5914.
10. Ibid., p. 7203.
11. Ibid., p. 7806.
12. Ibid., p. 7206.
13. Ibid., p. 6521.
14. *Senate Watergate Report*, pp. 596-97.
15. Ibid.
16. Ibid., pp. 598-99.

17. *Senate Watergate Hearings,* pp. 6524-32, 6693, 7628-29.
18. Ibid., pp. 6537, 6778.
19. Ibid., p. 5869.
20. (120-121H)
21. *Senate Watergate Hearings,* pp. 6329-30.
22. U.S. Congress, House Judiciary Committee, 93rd Cong., 2d Sess., 1974 *Statement of Information* (Hearings pursuant to H. Res. 803, A Resolution Authorizing and Directing the Committee on the Judiciary to Investigate Whether Sufficient Grounds Exist for the House of Representatives to Exercise Its Constitutional Power to Impeach Richard M. Nixon, President of the United States of America) (Washington, D.C.: U.S. Government Printing Office, 1974), p. 271 (hereinafter referred to as *Statement of Information*).
23. *Senate Watergate Report,* p. 792.
24. Ibid., p. 793.
25. *Senate Watergate Report,* p. 621; *Senate Watergate Hearings,* p. 6799.
26. *Senate Watergate Hearings,* p. 6263.
27. *Senate Watergate Report,* p. 785.

NOTES FOR CHAPTER SIX

1. *Senate Watergate Report,* p. 624.
2. *Senate Watergate Hearings,* pp. 6635, 6711, 7593.
3. Ibid., p. 385.
4. *Statement of Information,* pp. 366-67.
5. Ibid., p. 382.
6. *Senate Watergate Report,* pp. 6134-35; *Statement of Information,* p. 530.
7. Ibid., p. 6715.
8. *Statement of Information,* p. 518.
9. *Senate Watergate Report,* p. 635.
10. Ibid., p. 637.
11. *Statement of Information,* p. 396.
12. Ibid., pp. 520-21.
13. Ibid., pp. 570-71.
14. Ibid., pp. 574-75.
15. Ibid., pp. 578-80, 592-83, 586, 588.
16. Ibid., p. 598.
17. *Senate Watergate Hearings,* pp. 7067, 7141; *Senate Watergate Report,* p. 646.
18. *Senate Watergate Report,* pp. 631-35.
19. *Statement of Information,* p. 641.
20. Ibid., p. 670.

21. *Senate Watergate Hearings,* p. 6632.
22. Ibid., pp. 7074, 7075.
23. *Senate Watergate Report,* pp. 657-58.
24. *Statement of Information,* p. 176; *Senate Watergate Hearings,* pp. 6700, 7601, 7602, 7814.
25. *Senate Watergate Hearings,* pp. 7848-50.
26. *Senate Watergate Report,* pp. 668-70.
27. *Senate Watergate Hearings,* p. 8139.
28. *Ibid.,* pp. 8141-42.
29. *New York Times* (December 17, 1972), p. 61.
30. *New York Times* (January 11, 1973), p. 1.
31. *New York Times* (May 22, 1973), p. 29.
32. *New York Times* (June 11, 1973), p. 1.
33. *New York Times* (July 12, 1973), p. 27.
34. *New York Times* (July 28, 1973), p. 9.
35. *New York Times* (August 16, 1973), p. 38.
36. *New York Times* (September 13, 1973), p. 37.
37. *New York Times* (September 19, 1973), p. 36.
38. *New York Times* (November 13, 1973), p. 20.
39. *New York Times* (December 5, 1973), p. 30.
40. *New York Times* (December 20, 1973), p. 34.
41. "Text of Nixon Letter and Ervin Reply," *New York Times* (January 5, 1974), p. 15.
42. Ibid.
43. John Herbers, "Nixon Says He Considered Politics in Milk Rise But Denies Any Deal," *New York Times* (January 9, 1974), p. 1.
44. Phillip Shavecoff, "Milk Price 'White Paper' Seems to Contradict Nixon," *New York Times* (January 10, 1974), p. 21; *Watergate Special Prosecution Force Memorandum* "Hardin Prosecution," September 20, 1974.
45. *New York Times* (May 3, 1974), p. 16.
46. *New York Times* (May 3, 1974), p. 28.
47. "Excerpts from the News Conference by St. Clair," *New York Times* (May 8, 1974), p. 34.
48. William Robbins, "Judge Suspends Nader Milk Suit," *New York Times* (May 15, 1974), p. 31.
49. "St. Clair Letter on Milk," *New York Times* (May 23, 1974), p. 32.
50. James M. Naughton, "House Panel Warns Nixon It May Find Tape Refusal Ground for Impeachment," *New York Times* (May 31, 1974), p. 10.
51. "16 Impeachment Panel Members Got Election Aid from Dairymen," *New York Times* (June 5, 1974), p. 30.
52. *New York Times* (June 25, 1974), p. 22.
53. David E. Rosenbaum, "Kalmbach Says Dairymen Had to Reaffirm

Pledge," *New York Times* (July 18, 1974), p. 21.
54. *New York Times* (July 20, 1974), p. 15.
55. "Proposed Articles of Impeachment Submitted to Panel by Counsel," *New York Times* (July 20, 1974), p. 17.

NOTES FOR CHAPTER SEVEN

1. *United States of America v. John B. Connally*, United States District Court, District of Columbia, Criminal Action No. 74-440, Official Transcript of Proceedings, p. 1311.
2. James M. Naughton, "Connally Acquitted of Bribery Charge; Hints He May Resume Political Career," *New York Times* (April 18, 1975), p. 1.
3. Ibid.
4. Larry L. King, "Williams for the Defense—The Trial of John Connally," *Atlantic* (July 1975), p. 49.
5. Ibid.
6. Sandra Salmans and Stephan Lesher, "A Reputation Retrieved," *Newsweek* (April 28, 1975), p. 35.
7. Aaron Latham, "John Connally on the Comeback Road," *New York* (October 27, 1975), p. 48.
8. Ibid.
9. Ibid., p. 53.
10. Ibid.
11. Ibid.
12. Gail Cameron, "Nelly Connally: First Lady in Waiting?" *McCall's* (August 1973), p. 117.
13. Watergate Special Prosecution Force Memorandum dated June 18, 1974 to Henry S. Ruth, Jr. from Frank M. Tuerkheimer, Jon A. Sales and James L. Quarles, III re John B. Connally (June 18, 1974), p. 2.
14. Ibid., p. 3.
15. Ibid.; *Senate Watergate Report*, p. 638; *Senate Watergate Hearings*, p. 6407.
16. *Senate Watergate Hearings*, p. 6058.
17. Ibid.
18. Ibid.
19. Watergate Special Prosecution Force Memorandum re John B. Connally, p. 4.
20. King, p. 41.
21. *Senate Watergate Hearings*, p. 6421; Connally Trial Transcript, pp. 1152-53.
22. *Senate Watergate Report*, pp. 683-84; *Senate Watergate Hearings*, pp. 5961-62.
23. Ibid.

24. Watergate Special Prosecution Force Memorandum re John B. Connally, p. 6.
25. Ibid.
26. Ibid.
27. Ibid.
28. Ibid.
29. Ibid., pp. 6, 7.
30. Ibid., p. 7.
31. Ibid.
32. Watergate Special Prosecution Force Memorandum re John B. Connally, p. 7; Connally Trial Transcript, p. 1159.
33. Watergate Special Prosecution Force Memorandum re John B. Connally, p. 2.
34. Ibid., p. 8.
35. Ibid.
36. Ibid., p. 9.
37. Ibid., p. 17.
38. Ibid.
39. Ibid., pp. 9, 10.
40. Ibid., p. 16.
41. Ibid., pp. 16, 17.
42. Ibid., p. 18.
43. King, p. 46.
44. Ibid., p. 47.
45. Watergate Special Prosecution Force Memorandum re John B. Connally, pp. 10, 18.
46. Ibid., p. 14.
47. Ibid.
48. Watergate Special Prosecution Force Memorandum re John B. Connally, p. 18; King, p. 45.
49. Watergate Special Prosecution Force Memorandum re John B. Connally, pp. 10, 11; King, p. 42.
50. King, p. 45.
51. Ibid., p. 47.
52. Ibid., p. 46.
53. Watergate Special Prosecution Force Memorandum re John B. Connally, pp. 11, 12.
54. Watergate Special Prosecution Force Memorandum re John B. Connally, pp. 11, 12; King, p. 42.
55. Watergate Special Prosecution Force Memorandum re John B. Connally, pp. 11, 12; King, p. 43.
56. Watergate Special Prosecution Force Memorandum re John B. Connally, p. 12.
57. Watergate Special Prosecution Force Memorandum re John B. Connally, pp. 12, 19.
58. "Lawyer Is Indicted in Milk Payoff," *Cleveland Plain Dealer*

(February 22, 1974), LA Times/Washington Post Service, p. 10.

59. "Lawyer From Texas Pleads Not Guilty in Milk Case," *Cleveland Plain Dealer* (March 16, 1974, LA Times/Washington Post Service), p. 3B.
60. Jack Anderson, "Connally Target of Bribe Probe," *Cleveland Press* (April 10, 1974).
61. "New Bills Replace Dairy Co-Ops Left for Connally," *Cleveland Plain Dealer* (April 11, 1974), p. 1.
62. "Milk Aide Said to Accuse Connally of Taking Payoff," *Cleveland Plain Dealer* (April 20, 1974), p. 1.
63. "Connally Again Denies Allegation He Took Milk Payoff," *Cleveland Plain Dealer* (April 21, 1974), p. 10AA.
64. Ibid.
65. George F. Will, "That's All for a While," *National Review* (June 20, 1975), p. 522.
66. Scott Armstrong and John F. Berry, "Nixon Attempted, Unsuccessfully, to Halt Prosecution of Connally," *The Washington Post* (December 28, 1979), p. 1.
67. Brooks Jackson, "Plea Bargaining Arrangements to Get Testimony Against Connally," Associated Press (June 21, 1974).
68. "Lawyer from Texas Pleads Guilty to Bribing Connally," *Cleveland Plain Dealer* (August 8, 1974), p. 1.
69. Watergate Special Prosecution Force Memorandum re John B. Connally, p. 20.
70. Ibid., pp. 20, 21.
71. Watergate Special Prosecution Force Memorandum re John B. Connally, pp. 21, 22; *Senate Watergate Report*, pp. 686, 687.
72. Watergate Special Prosecution Force Memorandum re John B. Connally, p. 22.
73. Ibid., pp. 24, 25.
74. Ibid., p. 25.
75. Ibid.
76. Ibid.
77. Ibid., pp. 23, 24.
78. Ibid., p. 25.
79. Ibid., pp. 25, 26.
80. Ibid., p. 27.
81. Ibid.
82. Ibid., p. 28.
83. Ibid., p. 29.
84. Sandra Salmans and Stephen Lesher, "One Man's Word," *Newsweek* (April 21, 1975), p. 46.
85. King, p. 45.
86. Ibid.
87. Ibid.

88. Graham, p. 8.
89. Latham, p. 55.
90. Ibid., p. 56.
91. Ibid.
92. Ibid.
93. Ibid.
94. William Safire, "Cross Hairs on John Connally."
95. "Connally's Accuser Is Given Probation," *Cleveland Plain Dealer* (August 21, 1976, LA Times/Washington Post Service), p. 1.
96. Ibid.

NOTES FOR CHAPTER EIGHT

1. *Senate Watergate Report,* p. 624.
2. Ibid., p. 8022.
3. Ibid., pp. 7981-86.
4. *Senate Watergate Report,* pp. 700-701.
5. *Senate Watergate Hearings,* p. 7485.
6. Ibid., p. 7491.
7. Ibid., p. 7499.
8. *Senate Watergate Report,* p. 708.
9. *Senate Watergate Hearings,* p. 8153-54.
10. Ibid., p. 7983.
11. Ibid., p. 7680.
12. Ibid., p. 7984.
13. Ibid., p. 7985.
14. Ibid.
15. Ibid.
16. Ibid., p. 8005.
17. Ibid., pp. 7229, 7232-33.
18. Ibid., pp. 7232-33.
19. Ibid., pp. 6439-6444.
20. Ibid., pp. 6122-24.
21. Ibid., p. 6448.
22. Ibid., pp. 6444-45.
23. Ibid., p. 6453.
24. Ibid., pp. 7614-7615.
25. Ibid., p. 6117.
26. Ibid., p. 6118.
27. Ibid., p. 6118.
28. *Senate Watergate Report,* p. 726.
29. Ibid., p. 6121.
30. Ibid., p. 7615.
31. Watergate Special Prosecution Force Memorandum, July 19, 1974

re George L. Mehren; Watergate Special Prosecution Force Memorandum, September 27, 1974, re Mehren prosecution.
32. Ibid.
33. Ibid.
34. Watergate Special Prosecution Force Memorandum, September 27, 1974, re Mehren prosecution, p. 11.
35. Ibid., p. 13.
36. *Senate Watergate Report*, p. 729.
37. *United States v. Associated Milk Producers, Inc.*, U.S. District Court, Western District of Missouri, Western Division, Civil Action No. 74-CV-80-W-1.
38. November 1, 1974, letter to United States District Judge John W. Oliver from Paul W. Walter, Michael T. McMenamin and David R. Williams in Civil Action No. 74-CV-80-W-1.

NOTES FOR CHAPTER NINE

1. Brooks Jackson, "Milk Money," *New Republic* (August 10, 1974), p. 11.
2. "Milk Producers Really Butter Politicos," *Canton Repository* (February 18, 1973), p. 51.
3. Ibid.
4. Ibid.
5. Watergate Special Prosecution Force Memorandum re Stuart H. Russell, December 12, 1974.
6. Watergate Special Prosecution Force Memorandum re Jack Valentine and Norman Sherman, June 25, 1974.
7. Ibid.
8. Ibid.
9. Ibid.
10. Ibid.
11. Ibid.
12. Ibid.
13. Ibid.
14. *Report of Wright* et al., p. 100.
15. Frank Wright, "Top Milk Aide Tells of Repeated Bids to Raise $25,000 for Mondale Race," *Minneapolis Tribune* (May 18, 1974), p. 1.
16. Ibid.
17. Ibid.
18. Ibid.
19. Ibid.
20. Ibid.
21. Watergate Special Prosecution Force Memorandum re Stuart H. Russell, December 12, 1974.

22. Mitchell C. Lynch, "How Earl Butz Took Pen in Hand and Tried to Help Out a Friend," *Wall Street Journal* (September 17, 1973), p. 1.
23. Ibid.
24. Ibid.
25. Michael Barone, Grant Ujifusa, and Douglas Mathews, *Almanac of American Politics 1978*, pp. 695-96.
26. *Senate Watergate Report*, pp. 607-9.
27. Ibid.
28. Watergate Special Prosecution Force Memorandum re Harold Nelson, June 28, 1974.
29. *Senate Watergate Report*, pp. 607-609.
30. Watergate Special Prosecution Force Memorandum re Harold Nelson, June 28, 1974.
31. "Big Milk Co-Ops Gave $67,670 for Key Vote," *Cleveland Plain Dealer* (February 7, 1975), p. 8A.
32. Ibid.
33. Ibid.
34. "Milk Money Still Pouring into Political War Chest," *Cleveland Plain Dealer*, June 15, 1974.
35. Ibid.
36. Mary Meehan, "Rigging Election Reform," *Inquiry*, October 30, 1978, p. 16.
37. "Money's The Name of Politics' Game," *Madison, Wisconsin, Capitol Times* (November 8, 1978).
38. Food and Agriculture Act of 1977 (P.L. 95-113), p. 1898.

NOTES FOR CHAPTER TEN

1. William H. Nolte, "Portrait of a Civilized Man," *American Spectator,* (March 1979), p. 18.
2. Joe Western, "How the Public Got Milked," *National Observer* (April 2, 1977), p. 1.
3. *New York Times* (March 10, 1977), p. 43.
4. Western, p. 14.
5. Ibid.
6. Ibid.
7. Ann McFeatters, "Milk Support Likely to Cut Consumption," *Cleveland Press* (March 23, 1977).
8. "We're Being Milked," *Cleveland Press* (March 23, 1977), p. A6.
9. Emily Yoffe, "Two Milk Duds," *New Republic* (July 23, 1977), pp. 9-10.
10. James Grant, "Milk and Honey," *Barron's* (May 30, 1977), p. 7.

11. *Dairy Industry Newsletter* (April 12, 1978), p. 2.
12. Ibid., p. 3.
13. Ibid.
14. Ibid. (February 2, 1978), p. 2.
15. *Community Nutrition Institute, Proceedings of Conference on Milk Prices and the Market System* (Washington, D.C. 1975), pp. 140-141.
16. *Dairy Industry Newsletter* (March 3, 1977), p. 6.
17. Interview with Carol Foreman, June 1978.
18. Douglas Martin, "Carter's Just a Piker; Buster's Peanut Crop Covers 2,000 Acres," *Wall Street Journal* (June 3, 1976), p. 1.
19. Ibid.
20. Ibid.
21. Ibid.
22. "President Carter," *Congressional Quarterly* (Washington, D.C., 1977), p. 71.
23. *Dairy Industry Newsletter* (April 30, 1977), pp. 1-2.
24. Bill Curry, "Ice Cream Formula War: Dairymen in No Good Humor," *Washington Post* (June 4, 1977), p. A2.
25. *"President Carter,"* p. 71.
26. Nicholas Wade, "Ice Cream: Dairymen Imperiled by FDA's Recipe," *Science* (August 27, 1977), p. A44.
27. Ibid.
28. Ibid., A47.
29. *Dairy Industry Newsletter* (August 17, 1977), p. 1.
30. Ibid. (August 31, 1977), pp. 1-2.
31. Ibid. (October 26, 1977), p. 2.
32. Ibid. (December 21, 1977), pp. 1-2.
33. Ibid. (February 15, 1978), p. 2.
34. Ibid. (December 21, 1977), p. 2.

NOTES FOR CHAPTER ELEVEN

1. Les Whitten, *Conflict of Interest* (New York: Doubleday, 1976), p. 1.
2. Community Nutrition Institute, *Proceedings of Conference on Milk Prices in the Market Systems* (Washington, D.C., 1975), pp. 10-14.
3. Paul H. Weaver, "Unlocking the Gilded Cage of Regulation" *Fortune* (February, 1977), p. 188.
4. *Dairy Industry Newsletter* (March 2, 1977), p. 3.
5. *Dairy Industry Newsletter* (June 8, 1977), p. 8.
6. *Dairy Industry Newsletter* (August 30, 1978), p. 2.
7. *Report of National Commission for the Review of Antitrust Laws*

and Procedures (Washington, D.C.: U.S. Government Printing Office, 1979), p. 253.

8. Kenneth D. Naden, "Antitrust Threats Loom for Farmer Cooperatives," *American Agriculturist* (October 1978), p. 21.
9. *Report of the National Commission for the Review of Antitrust Laws and Procedures*, pp. 265-66.
10. *Proceedings of Conference on Milk Marketing* (Washington, D.C.: Community Nutrition Institute, 1976), pp. 101-4.
11. James Grant, "Milk and Honey," *Barron's* (May 30, 1977), p. 7.
12. Ibid.
13. Laurence J. Peter, *Peter's Quotations, Ideas for Our Time*, (New York: William Morrow, 1977), p. 213.
14. Ibid., p. 399.

Appendix

WATERGATE SPECIAL   DEPARTMENT OF JUSTICE
PROSECUTION FORCE
MEMORANDUM

TO   : Philip A. Lacovara   DATE: August 5, 1974
          Counsel to the Special Prosecutor
FROM   : Frank M. Tuerkheimer
          John A. Sale
          James L. Quarles III
SUBJECT   : Bribery Prosecution of Associated Milk Producers, Inc.

I. INTRODUCTION

As you know, by memorandum to the Deputy Special Prosecutor of June 28, we recommended that no prosecution be brought in connection with the March 25, 1971 price support decision. This recommendation was predicated on the convergence of two different ideas: the absence of viable defendants and a lack of clear criminality. In a footnote on page 23 we discussed in summary terms the question of whether the initial offer of $2 million constituted a bribe under Section 201

[DELETION: Conclusion of prosecuting attorney concerning decision whether or not to prosecute. Material also reflects on culpability of an individual who was not prosecuted.]

This memorandum is designed to discuss the question of the initial offer in greater detail with a view towards possible prosecution of AMPI based on the underlying facts.

II. THE FACTS

A. *Background:*
Beginning with the $100,000 payment to Kalmbach on August 2, 1969, dairy industry contacts with the Nixon Administration be-

came increasingly more frequent. As you know, the dairy industry is directly affected by a large number of governmental decisions including tariffs on imports, price supports, school lunch programs and so forth. As a result, the leaders of AMPI, Harold Nelson and David Parr found it essential to develop their contacts and maintain access to appropriate channels.

At the same time the contacts with the Administration were being developed, the TAPE program was getting underway. Organized in 1969 and modeled after COPE, TAPE began to accumulate vast amounts of money for purposes of political contributions. The collection of money involved a check-off system whereby each participating farmer paid $99 per year or ⅓ of 1% of his yearly gross receipts, whichever was less. Since AMPI's membership was about 40,000 people, and the membership of three main co-ops about 75,000 people, you can see that a system which involves collection of $100 per member per year leads to the collection of several million dollars.

B. *The Offer:*

On July 7, 1970, David Parr, the number two man in AMPI, met with Colson in Colson's office and said to Colson that the dairy industry was prepared to contribute $1- to $2 million to the President's re-election in 1972.*

Colson then wrote down the words $2 million on a piece of paper and then he and Parr agreed that the dairy industry would in fact make the $2 million contribution. Colson says that the commitment involved a promise by Parr to support no Democratic candidates; Parr denies this and says that his options were open.

[DELETION: Quotation from grand jury testimony.]

C. *Execution of the Commitment:*

Because of the surprising difficulty of getting committees organized to receive the funds, execution of the commitment was delayed by several months. TAPE contributed $135,000 to the 1970 Congressional effort of the Republicans and in March and April of 1971, the dairy industry contributed $85,000. In July of 1971 TAPE contributed another $125,000 and in the remainder of 1971, the other co-ops contributed $50,000. TAPE, in September of 1971, contributed an additional $62,500. In the closing days of the 1972 campaign, the other co-ops contributed an additional $75,000 to Democrats for Nixon and AMPI contributed $352,500, ostensibly to the Republican Senatorial and Congressional effort, but in fact to the President's

---

*Harold Nelson, AMPI's general manager, said the initial commitment was at least $2 million and perhaps more.

re-election campaign. Thus the total of the contributions is $885,000.

White House memoranda reflect the commitment as well. For example, a February 1, 1972 Strachan to Haldeman memorandum says that "Kalmbach is very concerned about his involvement in the milk producer's situation. He believes that Jacobsen and Nelson will deliver though they have cut the original 2,000 commitment back to 1,000." On February 16, 1972, Strachan noted to Haldeman that "Kalmbach is working with the milk people to increase the 233 currently banked to 1,000 by April 7." Thus, a substantial portion of the initial commitment reduced as it was in early 1972, was followed through on.

Perhaps the most blatant example of an attempt to link the commitment with governmental decision making can be found in a letter written by Patrick Hillings to President Nixon in December of 1970. Hillings, then an attorney for AMPI, was concerned with the delay surrounding the enactment of Tariff Commission recommendations. In that letter which was identified as "Tariff Commission (Milk) Recommendations", Hillings said the following: "We are now working with Tom Evans and Herb Kalmbach in setting up appropriate channels for AMPI to contribute $2 million for your re-election."*

Colson has testified that throughout the period following the commitment, the fact of the commitment was constantly on his mind. In an appearance before the House Judiciary Committee, he said:

"I don't think that there is any way that you can separate out of your mind the subconscious impact of somebody coming in and saying we are going to give $2 million . . . you can say all you want until your face turns blue that there are no *quid pro quos* but . . . there is no way that any man cannot be influenced and. . . . It is just unavoidable."

Colson, who acknowledged that he understood full well when the commitment was made that it was designed to influence otherwise close substantive decisions affecting milk producers, also made clear that Parr and Marion Harrison, an attorney for AMPI, constantly reminded those of his office of the fact and the size of the contribution. He testified before the Judiciary Committee that:

"The great difficulty with the milk producers was no matter whether you told them they weren't going to get any special consideration for their contributions or not, seemed to make no

---

*Colson says he intercepted the letter and that it never reached the president. He adds that he chastised the milk people for their crude tactics.

difference. They kept coming back and demanding things and say that after all, we are contributing all of this money."

Colson then added that after the repeated reminders "that they can't talk that way," Parr, Harrison and on the occasion of the December 18 letter, Patrick Hillings, would apologize only to revert to a similar form on a subsequent occasion.

"In defense" of Nelson, Parr, Harrison and Hillings, Colson said that they were only making explicit the implicit understandings with respect to any large contributor. In short, their crudeness lay not in expecting favorable treatment as a result of the large contribution, but making that expectation clear for all to hear.

III. RECOMMENDATIONS

A. *Statutes Involved:*
Section 202(b) of Title 18 provides that: "Whoever, directly or indirectly, corruptly ... offers or promises any public official ... to give anything of value to any other person or entity, with intent—(1) to influence any official act ... shall be fined not more than $20,000 or three times the monetary equivalent of the thing of value, whichever is more, or imprisoned for not more than 15 years, or both ... ."

B. *Analysis:*
It is our judgment that from a technical point of view, the requirements under Section 201(b) are met in this case.

[DELETION: Conclusion of prosecuting attorney concerning culpability of an individual who was not prosecuted for the violations described.]
[DELETION: Conclusion based on grand jury testimony.]

That those acts were undetermined at the time the offer or promise was made does not change the legal status as official acts. Thus, the elements under Section 201(b) are met.

C. *Subjects of the Charge:*
There are four legal entities directly involved in the violation: Nelson, Parr, Colson, and AMPI. Each will be discussed in turn.
1) Nelson—As you know, Nelson has already plead guilty to a charge alleging his authorizing approximately ⅓ of a million dollars in illegal corporate contributions, his authorization of a $100,000 payment to Kalmbach made for the purpose of securing access for AMPI officials to the White House and his authorization of payments to John Connally for his assistance on the March 25, 1971 price support decision.

[DELETION: Conclusions of the prosecuting attorney concerning the decision whether or not to prosecute. Material also reflects on culpability of non-indicted individual.]

[DELETION: Conclusion based on grand jury testimony. Material also reflects on internal deliberative process of prosecution.]

[DELETION: Refers to grand jury testimony.]

2) Parr—While the charge to which Parr has plead guilty is a weaker one than the Nelson charge, it still represents a significant plea; approximately $225,000 in illegal corporate contributions was acknowledged by Parr.

[DELETION: Conclusion of prosecuting attorney concerning the decision whether or not to prosecute and reflecting on culpability of a non-indicted individual.]

3) Colson—Under our plea-bargain with Colson, he is immune from prosecution for AMPI.

4) AMPI—The traditional test for criminal liability for the acts of its officers and agents is whether the acts of the officers and agents were done within the general scope of their authority and whether they were designed to benefit the corporation. It seems very clear that both these criteria are met by the facts of this case. Nelson and Parr were obviously authorized to act on behalf of AMPI in the areas of campaign contributions. If proof were lacking of such authorization, which it is not, then the subsequent contribution of several hundreds of thousands of dollars pursuant to the commitment firmly establishes Nelson's and Parr's authority to act. As for the requirement that the acts be for the benefit of the corporation, the evidence is even more compelling since Parr and Nelson had no direct interest in the decisions they were hoping to influence. Thus the corporation which has not been granted immunity for anything other than Section 610 violations is a proper subject of the charge.

5) Others—A number of other persons are involved in the execution of the $2 million commitment. The President's knowledge of it is proved by Colson's September 9, 1970 memorandum to him advising him explicitly of the $2 million commitment made by AMPI leaders to his re-election effort.

[DELETION: Segment identifies individuals who were involved in possible criminal activity but were not prosecuted.]

and perhaps several additional persons are all involved in one way or another in the process whereby the commitment was partially fulfilled. There are several reasons, however, why we feel that none of these persons would be the subject of a charge. First, the crime took place in July of 1970; the involvement of all others is subsequent to that. Second, while one may infer that all those involved in the execution of the commitment might have been aware of the intent which underlay it, the inference itself does not seem strong enough to take the case to the jury. Third, some of these other persons are immune. The evidence against others varies. In sum, for all these reasons, we think that the only serious thoughts of prosecution would be directed to the four entities directly involved, and there as has been urged, only AMPI is a reasonable subject.

IV. CONCLUSIONS

A. *Guilty Plea/Fine Problem:*

Erwin Heininger, attorney for AMPI, has indicated to us a willingness to plead guilty to some Section 201 charges in connection with AMPI's activities under Nelson's aegis. He has intimated that his authority is limited by a total amount of fines that might be imposed and that in turn a $100,000 is in the upper reaches of that decision. As you know, AMPI has already been fined $35,000.

Under Section 201(b), the maximum fine in this case is $6 million. There is no question but that AMPI would not plead guilty to a charge which would involve a fine of anywhere near that magnitude. We doubt whether Heininger would be able to go beyond his present $65,000 limitation.

Accordingly, we urge that pre-plea negotiations might be undertaken with Judge Hart in which his agreement to impose no more than $20-, $50- or $65,000 is obtained. Under these circumstances, we think that a plea could be arrived at.

B. *The Not Guilty Plea Problem:*

On the assumption that a guilty plea under any circumstances is out of the question, the issue arises whether we should proceed anyway. While our thinking has not crystallized on this issue since there is every indication that there will be a plea, we prefer to postpone the question until we know that there is a real one.

C. *Public Relations:*

In a sense, it may create public relations problems for the office in that we save only a corporation for the most serious charge of all.

[DELETION: Conclusion of prosecuting attorney concerning decision whether or not to prosecute. Material also reflects on culpability of persons not prosecuted.]

We feel, however, that the value of the prosecution itself, i.e. putting large contributors on notice that they are flirting with a Section 201 charge where the facts warrant is a highly desirable result and worth the inevitable criticism that will be incident to the prosecution of a corporation. For this reason, we recommend a charge as follows:

On or about July 7, 1970, in the District of Columbia, Associated Milk Producers, Inc. the defendant, through its officers and agents, unlawfully, willfully and knowingly, did directly and corruptly offer and promise to Charles W. Colson, Special Counsel to the President of the United States, a public official, to give $2,000,000.00, a thing of value, to the re-election campaign of President Richard M. Nixon, with intent to influence official acts, to wit, governmental decisions affecting the dairy industry.

(Title 18, United States Code, Section 201(b).)

[DELETION: Allegation of wrongdoing against individuals who were not prosecuted for this particular activity. Material also reflects on the internal deliberative process of prosecution force.]

Recommendation later changed by Tuerkheimer et al to "no prosecution"—no chance of a plea plus not convinced on the evidence, i.e., theory became that decision having been made, administration tried to capitalize on it by making AMPI fulfill its pledge. Original pledge not connected with price support decision particularly, i.e., just support generally. Also, charging only corporation and not individual penalizes only the farmer member of the AMPI who has to pay any fine. Individual defs found guilty and served jail time on other matters already.

HR*

WATERGATE SPECIAL   DEPARTMENT OF JUSTICE
PROSECUTION FORCE
MEMORANDUM

TO   : Henry S. Ruth, Jr.   DATE: June 18, 1974
              Deputy Special Prosecutor
FROM   : Frank M. Tuerkheimer
              John A. Sale
              James L. Quarles III
SUBJECT   : John B. Connally

I. INTRODUCTION

This memorandum is designed to set forth the facts which we can prove at a trial of John Connally, and analyze the factual and legal issues in the case. It is written in support of the proposed indictment which is attached to the memorandum.

---

*"HR" refers to Henry Ruth, the *third* Watergate Special Prosecutor.—Ed.

As you can see, the indictment is in eight counts. Jacobsen is named in Count 1 and charged with a violation of 18 U.S.C. § 201(f). As you know, he will plead guilty to the charge. Connally is charged with conspiracy to commit perjury and obstruct justice in Count 4 and with perjury in Counts 5, 6, 7 and 8.

Proof offered on the illegal payment counts (2 and 3) will be admissible on the obstruction counts (4 through 8) as evidence of motive. Proof offered on these obstruction counts will be admissible on the illegal payment counts as false exculpatory statements constituting admissions of the charge. Therefore, all the evidence referred to below will be pertinent to all of the charges against Connally.

II. THE PROSECUTION'S CASE

A. 1971 Events

The prosecution's case will rest on about twenty witnesses, almost all of whom are witnesses whose credibility will probably not be challenged. Three witnesses whose credibility may be in issue are Larry Temple, Bob Lilly and Sam Barnett. The credibility of Jake Jacobsen will very much be in issue.

Jacobsen, who is now 54, has been involved with leading Texas politicians since the late 1940's when he went to work for then Attorney General Price Daniels. He subsequently became a close aide of Daniels when Daniels was elected Senator and then Governor. From 1965 to 1967 he was a Special Legislative Assistant to President Johnson. He has known John Connally for the entirety of this 25 year period. Except for the time when Connally was Secretary of the Treasury, Jacobsen had virtually no opportunity for business contacts with him and as a result, their contacts over the 25 years involved mainly political matters.

The closeness of the Jacobsen/Connally relationship can be seen from Connally's Treasury logs. They show that in the period February 11 to June 30, 1971, the first three and one-half months of Connally's tenure as Secretary of the Treasury, Jacobsen met with him on ten separate occasions, more meetings than with any non-government person; indeed, more than twice as many meetings as with any of Connally's other non-Government callers. The logs further reveal that these meetings were not of a fleeting nature. The ten meetings last for over seven hours which is more than twice as much time as Connally spent with any other non-governmental person in his office during the three and one-half months.

After Jacobsen left the Johnson White House in 1967, he returned to Austin and began a law practice with Joe Long. This partnership remained in effect until June 1972 when Jacobsen filed a petition in bankruptcy. Associated Milk Producers, Inc. became one of the firm's

clients with the understanding that it was to pay $2,500 a month plus
whatever overages were billed. Pursuant to this arrangement, between
January 1, 1969 until April, 1972, Jacobsen and Long were paid
$100,000 in retainer fees plus an additional $147,000 in overages.
Much of the work which Jacobsen did for the AMPI account involved
speeches around the country to farmer groups designed to acquaint
them with the political advantages of participation in the TAPE program. As you know, TAPE was the political arm of AMPI and provided the vehicle through which the funding of its enormous political
war chest was accomplished. The firm of Jacobsen & Long also did
local work for AMPI in the state of Texas.

In April of 1969, Jacobsen formed the law firm of Semer, White &
Jacobsen in Washington. He remained a partner in this firm until
his bankruptcy in June of 1972. AMPI was also a client of Semer,
White & Jacobsen from April, 1969 until June, 1972. During this
period, $95,000 was billed and paid pursuant to the monthly retainer
and $21,500 under an overages arrangement. In all cases, both with
Semer, White, and Jacobsen and Jacobsen and Long, these bills to
AMPI—totalling over $350,000.00—were without supporting data and
were paid almost by return mail.

In early 1971, the principal issue concerning AMPI was the milk
price support question. After Connally became Secretary of the
Treasury, Harold Nelson, the general manager of AMPI, who knew
of Jacobsen's close and long time association with Connally, asked
him to intervene with Connally to see what he could do to obtain an
increase in the price support level of $4.66 then in effect. At this
time, Nelson told Jacobsen that the dairy industry had previously made
a $2 million commitment towards President Nixon's re-election. Jacob-
sen places this conversation at the Madison Hotel in early March of
1971. Nelson agrees. Madison Hotel records show that both registered
at the Madison during the first week in March.

Jacobsen met with Connally for an hour in Connally's office in the
Treasury building before the March 12 decision. Both Connally's logs
and his grand jury testimony reflect this meeting and reveal the date
to be March 4, 1971. At the meeting, Jacobsen discussed the merits
of the price support question with Connally, mentioned the political
power of the dairy lobby as a consequence of its numbers and vast
financial resources, described for Connally the size of the three co-ops
and their check-off system which required participating farmers to pay
$99 per year or ⅓ of 1% of their yearly gross receipts, whichever was
less, to TAPE. Jacobsen also told Connally that the dairy lobby had
pledged $2 million to President Nixon's re-election. After he had
told Connally that he had heard that Agriculture and OMB were leading the fight against an increase, he asked Connally to intercede on
behalf of his client. Connally said he would see what he could do.

On March 12, 1971, Secretary Hardin announced that the price support level would remain at $4.66. Nelson called Jacobsen and asked him if he would see Connally again. Jacobsen agreed. Thereafter, Jacobsen and Nelson met again at the Madison Hotel and among other things, reaffirmed the idea that Jacobsen would again speak to Connally. At one of the two meetings with Nelson, Jacobsen said it would be helpful if AMPI would pledge some additional funds, the credit for the pledge to go to Connally. He did not, however, discuss this with Connally. Nothing ever came of it.

Between the March 12 and March 25 discussions, Jacobsen again met Connally in Connally's office. Connally's logs and his grand jury testimony reflect this meeting as well and fix the date as March 19. Jacobsen told Connally that his clients were unsatisfied with the March 12 decision and that they were making every effort at having it reversed. He also told Connally that if the March 12 decision remained in effect, the physical and fiscal support previously pledged to the Administration would be terminated; once again Connally said he would see what he could do.

On March 23, 1971, in the early morning, Connally called the President. White House memoranda refer to the call as one in which Connally urged the President to raise the price support level to $4.92. At a late afternoon meeting on March 23, the President, Connally and others met to discuss the problem. The tape recording of that meeting reveals that the subject of the price support issue was raised by the President when he asked Connally to tell the group what he had told the President earlier that day over the phone.*

The next ten or fifteen minutes of the meeting can fairly be characterized as an intense effort by Connally directed towards eliminating all options for the President other than to raise the price support level. Connally explicitly eschews dealing with the merits of the issue and spends a great deal of time discussing the political power of AMPI and the political consequences of any decision other than an increase. Connally mentions the three co-ops by name; he mentions the number of members each has; he discusses the check-off system, referring specifically to the $99 a year or ⅓ of 1% of the gross yearly receipts, whichever is less, arrangement. He tells the President that if the price support level is increased by legislation, then it will be the same drain on the budget as if it were increased administratively, except that Democratic leaders of Congress will get the credit for helping the farmers. Connally says that because of the dairy lobby's considerable financial resources and because of the significant legislators behind an increase, i.e., Carl Albert, Wilbur Mills, he thinks such a bill will

---

*The White House tape of the President's half of the morning telephone conversation indicates that Connally urged an increase.

pass. He then tells the President that a veto of the bill will cost him the electoral votes of three and probably six states in the farm belt. Immediately thereafter, the President agrees that he has no choice but to go up. He then says that he will.

[DELETION: Summaries of or references to grand jury testimony.]

Connally's logs confirm a phone conversation with Jacobsen on March 23, 1971. Jacobsen thereafter called Nelson and relayed the information onto him.*

On March 25, 1971, Secretary Hardin announced that the price support level would be set at $4.93. Agriculture/White House memoranda reveal that the decision was finalized the day before.

Between March 19 and April 28, Connally's logs show two phone calls and one meeting with Jacobsen. During this period Connally told Jacobsen that he had been of assistance on the price support issue and, in a meeting in Connally's office, about a month after the March 25 decision, Connally said that he had heard that the dairy lobby was being very generous with its money with the politicians and that Connally would appreciate it if Jacobsen would get some for him.**

Connally's logs show an April 28 meeting with Jacobsen. Lilly will testify and his notes show a long distance call on April 28 from Jacobsen asking for $10,000 "for John Connally's lock box" for his help on the price supports. Lilly will further testify that on May 4, 1971, he obtained the $10,000 which he gave to Jacobsen in Jacobsen's office on that day. Prior to obtaining the money, Lilly obtained the necessary authorization from Nelson who now confirms this specifically included an acknowledgement that, as he understood it, the money was to go to Connally personally.

About a week after the money was solicited by Connally, Jacobsen received $10,000 from Lilly in Jacobsen's law office in Austin. He put the $10,000 into safe deposit box #865 at the Citizens' National Bank in Austin the day he received it. A few days later, Jacobsen spoke

---

*Further evidence of Connally's advance knowledge of the decision is provided by Bob Lilly.

[DELETION: Summaries of or references to grand jury testimony.]

**By April 16, 1971, over $100,000 had been contributed by dairy interests to the president's re-election campaign.

Jacobsen says that no amounts were discussed but that he immediately fixed on the sum of $10,000 as the appropriate amount. He thereafter called Bob Lilly of AMPI and asked him if he could get $10,000 together for John Connally who had been helpful on the price supports and Lilly agreed to do it.

to Connally on the phone and told him he had what they had spoken about and he was ready to bring it and Connally said that was fine. A few days after that, Jacobsen took the $10,000 out of his safe deposit box and brought it with him to Washington. Bank entry records for box #865 reveal that at 4:50 p.m. on May 4, Jacobsen entered the box and that at 11:20 a.m. on May 13, 1971, he once again entered the box. Records at the Madison Hotel in Washington reveal that Jacobsen checked into the Madison at 9:45 p.m. on May 13. Connally's logs show two phone conversations with Jacobsen; one on May 7 and one on May 8, 1971. According to Connally and his secretaries, calls merely setting up appointments would not be reflected in the logs, only actual conversations with Connally.

On May 14, 1971, Jacobsen went to Connally's office and gave him $5,000, five bundles of ten $100 bills, each bundle separately wrapped, saying "this is part of what we talked about," whereupon Connally took the money to an adjoining bathroom and emerged a short time later without any sign of the bills. Later that morning Jacobsen opened a safe deposit box at the 15th and M Street branch of the American Security and Trust Company where he put the remaining $5,000. He is certain as to the date because of safe deposit box records.

Connally's logs show a one hour meeting with Jacobsen on May 14, from 10:15 a.m. to 11:15 a.m.* American Security and Trust Company records show that Jacobsen opened the box, entered at 11:42 a.m. and left at 11:47 a.m. Madison Hotel records reveal that he checked out at 2:04 p.m. that same day, May 14.

Jacobsen did not come to Washington during July or August of 1971. Connally's logs reveal one telephone call to Jacobsen in each of those two months. During the last four months of 1971, there are only four meetings between Jacobsen and Connally according to the Connally logs. Two of them, meetings of September 23 and 24, 1971, play a significant role in the second payment.

On September 23, Connally's logs show a meeting with Jacobsen from 2:25 to 3:10 p.m. The fact of this meeting can be found both in Connally's appointment calendar and his Treasury logs. According to Connally's secretary, any appointment set in advance through her is reflected on the appointment calendar. Meetings which Connally in fact had with persons are reflected on the Treasury logs, as is the precise time of the meeting. Madison Hotel records show that Jacobsen checked in at 1:00 p.m. on September 23.

During the September 23 meeting, Jacobsen asked Connally if he was ready for the rest of the money and when Connally said he was,

---

*The diary of Bill Camp, Controller of the Currency, reveals a Jacobsen visit from 11:15 to 11:30 on May 14. Camp's office is also in the Treasury building.

Jacobsen said he would bring the money over the next day. The arrangement was that Jacobsen would come by the next morning.

The next day, according to Jacobsen, he first emptied the contents of his safe deposit box at the American Security and Trust Company, brought the $5,000 to Connally's office, waited for Connally, and then gave him, in the same form as before, $5,000. Once again Connally went into the bathroom with the money and reappeared without any sign of it.

Records for the American Security and Trust Company show that Jacobsen entered and left his box at 9:20 and 9:21 a.m., respectively, on September 24, 1971. Bank records reveal no further entry records for this box and when it was examined on April 15, 1974, it was empty.

Connally's Treasury logs for September 24 show that at 9:00 a.m., Connally attended a cabinet meeting, that from 9:30 to 10:30 he met with the President and Arthur Burns and that at 11:00 he appeared at the Senate Foreign Relations Committee. The logs also show a meeting between Jacobsen and Connally from 10:35 until 10:45.* This meeting is not reflected in Connally's appointment calendar, suggesting very strongly it was set up during the meeting on September 23.**

Connally does not deny the story outright. He says he was offered the money twice, once at a lunch with Jacobsen on June 25, 1971, and once after Democrats for Nixon was formed in August, 1972, and he declined it on each occasion. On both occasions, he was, according to his testimony, alone with Jacobsen. Since the theory of the prosecution is that such testimony is pursuant to a conspiracy to commit perjury and obstruct justice, it is discussed most appropriately with other facts pertaining to this conspiracy. What is significant for these purposes, however, is Connally's inability to explain what the brief ten minute meeting was about on September 24 after the data on his logs was explained to him and that his defense precludes him from arguing that Jacobsen converted the money to his own use.

B. *The Obstruction Period*

Sometime in late October, 1973, Jacobsen received a call from Nelson saying that Lilly was prepared to tell investigators in Washington about money Lilly gave to Jacobsen for Connally in 1971. Thereafter, Jacobsen called Connally and told him that he had heard that the investigation in Washington was zeroing in on money which Jacobsen had gotten for Connally. Jacobsen then agreed with Connally's assertion that Jacobsen never gave Connally any of the money.

---

*The entries in Connally's Treasury logs reflect actual times visitors spent with the Secretary, not waiting time.

**Camp's records show he was out of town on September 24.

The rest of the call was spent with Connally telling Jacobsen how he was being investigated all over the place.

Jacobsen's toll records show a fifteen minute person-to-person call to Connally on October 24, 1973. These records refresh Jacobsen's recollection, telling him that both the Nelson and Connally calls were on October 24, 1973.*

Jacobsen next spoke to Connally between 9:00 and 10:00 a.m. on Friday, October 26 at the Sheraton-Crest Hotel in Austin. The meeting was arranged either at the end of the October 24 call or during a phone conversation between October 24 and the time of the meeting. At this meeting, Jacobsen told Connally he was not going to tell any-one that he gave him the money and Connally said this was fine. After discussing this at length, they talked about how the Lilly problem would be handled. Connally suggested that he would replace the $10,000, the amount Jacobsen thought was involved, and that Jacobsen would put it in his box and say it was there all the time. They next discussed how they would explain the money's retention for a three and one-half year period. Connally suggested that Jacobsen could say it was offered to him to give to other candidates while he was Secretary of the Treasury and that he refused to accept it because of his anomolous position as a Democrat in a Republican Administration. Connally also suggested that Jacobsen could say he re-offered him the money after Democrats-for-Nixon was formed but that it was declined then because of various AMPI problems. Finally, Connally suggested that controversy around Watergate could explain the further delay in returning the money.

Connally's itineraries show that he and Mrs. Connally did not arrive in Austin until 10:00 a.m. on Friday, October 26, and he has testified that they did not arrive until the mid-morning of October 26. The records from the Sheraton-Crest Hotel, however, show that they arrived on the evening of October 25 and

[DELETION: Summaries of or references to grand jury testimony.]

Finally, Jacobsen has told us that he recalled coffee being served during his hour stay with Connally. Hotel records reveal that Connally did in fact order coffee for three on October 26 and they identify the waiter as Sam.

[DELETION: Summaries of or references to grand jury testimony.]
[DELETION: Summaries of or references to grand jury testimony.]

---

*Nelson's toll records show a call to Jacobsen's number on October 24, 1973, as well.

After receiving the subpoena, Jacobsen called Connally at the Sheraton-Crest and when Connally was not in, left a message asking Connally to call him back.

[DELETION: Summaries of or references to grand jury testimony.]

toll records from Connally's home phone reveal a long distance call to Jacobsen on Sunday, October 28.

On October 29, Jacobsen arranged for the availability of a private plane and then called Connally's secretary to set up an 11:00 a.m. appointment. He arrived at Connally's office shortly before 11:00 and once again went over with Connally the same ground covered two days earlier at the Sheraton-Crest. In addition, Jacobsen and Connally discussed a reason Jacobsen could give for coming to see Connally that day and Connally suggested that the reason be that he and Jacobsen were discussing delay in the processing of a bank application of a Connally client named Gus Wortham. Connally then left the room and returned in about 10 minutes with a cigar box with $10,000 in it which he gave to Jacobsen. He said "this money should be all right, that it was all old enough." Jacobsen then left, went directly back to Austin, and put the money into safe deposit box #865 at Citizens' National Bank.

Records at Ragsdale Aviation show that Jacobsen arranged for a plane at 8:00 a.m. on October 29; that the plane left Austin at 9:30, arrived in Houston 10:25; that the plane left Houston at 12:35 and arrived back in Austin at 1:30. Access records to Jacobsen's safe deposit box #865 show an entry at 2:00 p.m. on October 29, 1973.

On November 2, 1973, Jacobsen testified before the grand jury.

[DELETION: Summaries of or references to grand jury testimony.]

Shortly after his return to Austin, Jacobsen was contacted by Larry Temple, an attorney and associate mutual to both Connally and himself. Temple suggested a meeting in his office with Marvin Collie, a partner in Connally's law firm, for November 9 and Jacobsen agreed. He met Collie at Temple's office on November 9 and was completely debriefed. He told Collie, among other things, about questions concerning the $10,000, his answers, and that he had consented to an inventory of the money. Jacobsen was told that Connally was due back on November 12 and would testify later in the week. The fact of this meeting and its general content is confirmed by Temple.

Jacobsen's toll records show a two-minute person-to-person call to Connally's number in Houston on November 12, 1973.

[DELETION: Summaries of or references to grand jury testimony.]

On November 14, 1973, Connally testified before the grand jury. His testimony also was consistent with the agreement of October 26 and in all material respects,

[DELETION: Identifies an individual who testified before the grand jury.]*
[DELETION: Summaries of grand jury testimony.]

When he got to Christian's house, Mr. and Mrs. Christian were there with some of their children, but Connally was not there. Shortly after, Mrs. Christian left to play tennis and then Connally arrived. He brought with him an attache case. After the three men talked, primarily about the investigation in Washington, Christian left the room and Jacobsen was alone with Connally.

Connally told Jacobsen that some of the money in the first batch of money contained Shultz bills and so were later than they should be and he had brought another $10,000 to replace that batch. When Jacobsen said he couldn't sign it, because of his promise to McBride and Sale, Connally told him he could work it out. Connally started to open his attache case and Jacobsen saw a bundle wrapped in newspaper inside, but then Connally suggested doing it outside. They said good-bye to Christian and went to Connally's car together. In the car, Connally gave him a second $10,000 and told Jacobsen he should keep the other $10,000 until the matter was over. Jacobsen then took the money home.

The November 24 call and November 25 meeting are corroborated by George Christian and Connally.

[DELETION: Summaries of grand jury testimony.]

The next day Jacobsen asked his former partner Joe Long to help him get to his safe deposit box without a record being made of it. Later that day he and Long went to the bank, Long got the master key and the two went into the vault together where Long took the contents of box #865 and gave it to Jacobsen, and put a package which Jacobsen gave him into Jacobsen's other safe deposit box which Long also opened, box #998.

[DELETION: Summaries of grand jury testimony.]
[DELETION: Summaries of grand jury testimony.]
[DELETION: Summaries of grand jury testimony.]

Virginia Straughn will identify her initials on Long's safe deposit access records for box #555 at 4:30 p.m. on November 26, 1973,

---

*Connally testified in basically the same way before the Ervin Committee on November 15, 1973.

when Long and Jacobsen came by. She will testify that she was away from her desk momentarily and when she returned, she found the two of them in the safe deposit vault. She told Long to sign his records which he did. She has no knowledge of what was going on in the vault, but can testify that Jacobsen was in a position to get to his box or boxes without signing in because he was in the vault with Long who had the master key.

The following day, November 27, Jacobsen received a call from the F.B.I. and thereafter, on the same day, he met two F.B.I. agents at the Citizens' National Bank where the $10,000 in box #998 was inventoried.* A witness from the Bureau of Engraving and Printing will testify that approximately fifteen of the bills that were inventoried on November 27, 1973, were not shipped from the vaults of the Bureau of Engraving and Printing until after June 1 of 1971. Several of these were not shipped until October of 1971. While we are also in a position to prove that as many as thirty additional bills were not in circulation as of June 1, 1971, we see no reason presently why such detailed proof is required. We intend only to show that one of these bills was not available for circulation to the public until March, 1973.

Sometime after the inventory Jacobsen received a call from Larry Temple, informing him that Temple's secretary was bringing an envelope to him. Soon thereafter Jacobsen received an envelope with the return address of Connally's law firm on it. A transcript of Connally's appearance before the Watergate Special Prosecution Force, a transcript of his appearance before the Ervin Committee, and a very detailed digest of Connally's grand jury testimony were in the envelope. Jacobsen will identify the envelope and its contents, which he read, and they will be put into evidence. Temple will testify that he picked the envelope up at Connally's law firm on Friday, December 7, 1973, with instructions to make it available to Jacobsen. Temple will further testify that while he glanced at the contents of the envelope, he did not read it, he did not keep a copy for himself and he did not discuss its contents with anyone. He simply had it given to Jacobsen on December 10, 1973.

[DELETION: Summaries of grand jury testimony.]**

On the eve of Jacobsen's indictment he was called by his attorney who told him the indictment would be filed the next day. Jacobsen

---

*The day after his indictment Jacobsen consented to the seizure of the $10,000. The money seized was the same as the money inventoried so it will be put into evidence.

**He had previously testified before the Ervin Committee on December 14, 1973, also in basic accord with the October 26 agreement.

thereupon told Joe Long, persons in his family, and Larry Temple. Temple will testify that he immediately called Connally, that Connally was busy but he told Connally's secretary it was important whereupon she put him through to Connally. Temple then told Connally he had just spoken to Jacobsen who said that he would be indicted the next day. Connally asked if he had any further information; Temple said he did not, and Connally thanked him.

Finally, Jacobsen will testify that the money which he received from Connally on October 29 was given to the F.B.I. in our office on March 7, 1974. F.B.I. agents will testify as to an inventory they prepared identfiying the $10,000. Examination of the $10,000, 280 bills, reveals that there are no bills signed by Connally as Secretary of the Treasury, but 49 bills signed by George Shultz as Secretary of the Treasury. We have pursued all of these 280 bills to see whether they were in circulation on October 29, 1973. Our investigation is almost complete and it shows that all the bills were available for circulation by then. This evidence is particularly instructive as far as the Shultz bills are concerned since many Federal Reserve Banks did not receive any Shultz bills until early 1973.

Since we are charging a conspiracy to commit perjury, it is, of course, necessary that we place before the petit jury not only the pertinent parts of Jacobsen's November 2, 1973 and January 25, 1974 grand jury testimony, but portions of Connally's grand jury testimony as well. We propose to put into evidence all of Connally's grand jury testimony which deals with the $10,000 as well as the entirety of his testimony dealing with events in the fall of 1973.*

With respect to the $10,000, Connally testified on November 14, 1973, that Jacobsen offered him $10,000 to be made available to Connally for him to send to any committee or candidate of his choice. Connally declined on the ground that as a Democrat in a Republican Administration he did not want to be involved. When questioned about the source of the $10,000, the following occurred:

"Q. Did Mr. Jacobsen tell you the source of this $10,000?
A. No, he did not.
Q. He didn't tell you it had come from AMPI or the dairy industry?
A. No."

However, when he testified on April 11, 1974, and was asked what Jacobsen told him when the offer was made, Connally said: "My best recollection is that he said that the milk producers were going to start making some political contributions in 1971 and that they would

---

*Parallel Ervin Committee testimony will be put in as well.

be contributing to candidates in both parties and that there was
$10,000 available then to be given to any committee or candidate or
campaign that I would designate."

"Q. So it is correct to say that when he first mentioned the
$10,000 to you, he did mention the source of the money.
A. Yes. I don't recall that he mentioned anything in particular.
As I recall, he did say milk producers."

Connally has, from the beginning, had a hard time in formulating
his story. After having said on November 14 that Jacobsen did not
mention the source of the funds, he testified:

"... in the discussion he pointed out that, and this I think
clearly indicated that it probably was milk money—he said that
they were going to try to make some of their contributions in
1971 as well as in 1972, and that he had available this $10,000 to
be placed as I would like for it to be designated, and I told him
that I had no interest in that at all and that was about the extent
of it."

Of course, there is nothing in what was said indicating "that it probably
was milk money"; Connally simply realized he made a mistake and
tried to correct it.

Connally also testified that Jacobsen renewed the offer of the
$10,000 shortly after Democrats for Nixon was formed in August of
1972. Democrats for Nixon was formed on August 9, 1972 and Connally said it was about then that the offer was made. He said it was
declined for three reasons:

1) AMPI had tax problems,
2) AMPI was a defendant in án antitrust suit, and
3) AMPI had internal problems.

When Connally was asked why in the list of reasons he gave for
rejecting the offer in August of 1972, he did not mention the unfavorable publicity AMPI had received in connection with the price
support decision and subsequent contributions, he said there was no
reason. When it was suggested to him that since the other co-ops and
their political arms had also received unfavorable publicity, but that
he did accept contributions from them, he said that "well ... we
thought about it for a considerable length of time and discussed it
within the framework of the structure that we then had as to whether
or not we ought to take any contributions from any co-ops." According to GAO reports, SPACE, the political arm of Dairymen, Inc.,
contributed $25,000 to Democrats-for-Nixon on August 10, 1972, one
day after it was formed.

With respect to the events of the fall of 1973, we have two distinct versions by Connally, each of which is markedly inconsistent with the facts described above.

On November 14, Connally, after having been asked about the $10,000, was asked:

"Q. When have you last discussed this matter with Mr. Jacobsen?
A. Oh, gosh, a long time ago. I don't recall."
He was then asked:
"Q. Have you discussed it with him recently, within the last three or four weeks?
A. No."

When he was next asked whether he had any conversations with Jacobsen in the last three or four weeks, Connally referred to a discussion about a bank application two and one-half or three weeks ago "but that is the only contact I have had with him." Connally then very briefly described the conversation as one in which a client of the firm's wanted to find out the status of a bank charter application.

Connally was then asked whether he discussed the $10,000 with Jacobsen during that conversation. When he said he did not recall, it was pointed out to him that given the peculiar nature of the $10,000 transaction, it is something he would recall if he discussed it with Jacobsen recently. Connally agreed that was probably right, and he had discussed the dairy thing with Jacobsen, but he did not recall whether it was done on that occasion and that whenever it was discussed, "we simply treated it just like we treat anything else, we are both going to tell the truth about it and that is all there is to it." When he was asked the last occasion on which he discussed the "dairy thing" with Jacobsen, he said "I don't recall having any major discussion with him since last fall."

On April 11, Connally's version of the events in the fall was markedly different from his November 14 testimony, both still in sharp contrast to facts outlined above. Connally was reminded that he testified about meeting Jacobsen in Houston on a date which his appointment calendar showed as October 29. He was advised that his lawyers had told us that the meeting was set up the day before when he called Jacobsen from his home. He was then asked what triggered the call to Jacobsen on Sunday, October 28. Connally said that he was taking care of a number of last minute matters before leaving for a two week trip the next day and it occurred to him that he had done nothing about a request from Mr. Gus Wortham, one of the firm's oldest and most important clients, made almost three weeks before to see whether anything unusual was delaying a bank charter application. Connally testified that the meeting with Wortham was on October 10, 1972. He said that because of the Agnew resignation on October 10 and all the ensuing calls he had forgotten about it and he did not think of it until October 28 when he called Jacobsen and asked him to come over the next day with one of the Ragsdale planes if necessary. When asked whether Jacobsen told him over the phone that he had been summoned to appear before the grand jury, Connally answered, "No, he did not then, he had already told me that." This prior meeting will be discussed below.

Connally agrees that Jacobsen called on the morning of October 29 to confirm that he was coming.

[DELETION: Summaries of grand jury testimony.]

Connally agreed that it was a "very very busy day." When he was asked why, because it was so busy, he did not handle the matter with Jacobsen over the phone, he answered "Well, yes, I think *the reason* I did not is it would have taken me longer, I think, on the telephone to do it than it would have taken in person.... (Emphasis added)

Connally then testified that Jacobsen got there about 10:45 and they spent 20 minutes together during which Connally described the bank application matter that he wanted Jacobsen to look into. He said also that he gave Jacobsen two $100 bills to pay for the flight but that he did not give him any files or folders in connection with the bank application. He also testified that he had not yet billed the client for the $200 he gave Jacobsen, but that eventually he would bill the client or the firm.

Gus Wortham will testify that he initially discussed the question of any unusually caused delay in his bank charter application with Connally at a meeting in Connally's office on August 20, 1973, that it was mentioned again on October 10, and that Connally had not provided an answer for him as late as April of 1974.

When Connally was asked whether Jacobsen told him over the phone on October 28 that he had been subpoenaed to appear before the grand jury, Connally answered "No, he did not then, he had already told me that." Connally then testified that on Friday, October 26, when he returned to his hotel room at about 5:00, he had a message to call Jacobsen; he called him and Jacobsen said he needed to talk to him and Connally told him to come down immediately. He got there, acording to Connally, about 5:20 and spent fifteen to twenty minutes talking. Mrs. Connally was in the vicinity, but did not participate in the conversation.

Hotel records and other extrinsic evidence do not support Connally. Records for Connally's stay at the Sheraton-Crest reflect no local calls. The General Manager of the Sheraton-Crest will testify that local calls are posted to the guest's record by a night auditor.

Calls made between a night auditor posting and check-out time are added at check-out time. He will further testify that this procedure is automatic and that all customers are charged for all calls.

Connally testified that in the late afternoon October 26 meeting at the Sheraton-Crest Jacobsen told him he had been subpoenaed and that Connally's name would be involved. Connally asked why his name was to be involved, and Jacobsen explained that Lilly had mentioned Connally's name in connection with conversations about possible contributions. Jacobsen then said, according to Connally, " 'And I am prepared to testify that our conversations never took place and that I never heard of any $10,000.' " Connally said that he then told Jacobsen that he could not do that, that since the conversations did take place, they would have to be testified to and that although Connally was not happy about it, he said it was just one of those things.

Connally then testified that this contact was the only contact he had had with Jacobsen until "you get way back up into the spring of 1973." He was then read that portion of his November grand jury testimony in which he was asked whether he had any conversations in the three or four weeks preceding November 14 which he answered by saying that he met him on the bank application matter "but that is the only contact I have had with him." Connally's answer was "Yes, I responded to that question in the context of what preceded it, asking me about the discussions with respect to money, and I responded, as I recall, that he had not talked about that, somewhere in here, that I had not talked about that since 1972, as I recall. I do not see it right here." Of course, the question was preceded by Connally's statement about "Oh, gosh, it was a long time ago" that he had spoken to Jacobsen about the $10,000, so there is nothing in the context which removes the inconsistency.

Connally was told of

[DELETION: Identifies individual who testified before the grand jury.]

Connally said he did not set it up and that he did not know what Jacobsen said at the debriefing. When asked if he was ever informed of what Jacobsen said before the grand jury, he said that "No, as a matter of fact, I got back on the 12th, as I recall, and appeared on the 14th, and I really don't know to this good day what Mr. Jacobsen, when he appeared before the committee or this grand jury or how many times or what he said."

Connally also testified about the meeting in George Christian's house on November 25, 1973. He said that he called Christian on November 24, asking Christian to set up a meeting with Jacobsen for the next morning so that they could talk about what to do about the Ervin Committee hearings if they resumed. He testified he arrived at Christian's house around 11:00 the next day, having come from his daughter's house where he intended to return to continue his visit with her and his grandchildren. Jacobsen was already there when he arrived. He said that he, Jacobsen was already there when he arrived. He said that he, Jacobsen and Christian talked about possible Ervin Committee developments, when Christian took one of his children who was running through the living room upstairs to put some clothes on him. Christian was gone, according to Connally, five or six minutes during which time he was alone with Jacobsen. Connally also stated that Jacobsen said very little during the entire meeting.

According to Connally, when he and Jacobsen left, he walked towards Jacobsen's car with him, and asked Jacobsen if he had checked on the Gus Wortham matter. Jacobsen answered in a way that suggested he had done nothing and Connally lectured him on the importance of it and then added that he told Jacobsen not to let his bankruptcy get him down and to straighten up and carry on. When he was asked whether he observed from Jacobsen's demeanor on November 25 that Jacobsen was depressed, Connally answered "Well, I observed it before that and that was one of the reasons I asked him to come to Houston on October 29, so I could personally look at him and impress upon him the seriousness and interest I had in this bank charter and to judge for myself whether or not he was going to be aggressive enough and outgoing enough..."

"Q. That is one of the reasons why you wanted to see him on the 29th rather than deal on the telephone?
A. That is one of them, yes."

When it was pointed out to Connally that according to his testimony he had seen Jacobsen on October 26, Connally said that "the meeting on the 26th was an extraordinary meeting in terms of the subject matter...[and]...I did not pay any attention to what his demeanor was or what his attitude was or anything else." Of course, Connally had testified that when Jacobsen told him he was willing to lie about the underlying events, "it occurred to me that he might, indeed, be saying that to see what reaction he would get from me, so I did not draw any harsh conclusions about it all."

Connally also testified that he had a briefcase with him when he went to Christian's house on November 25, 1973.

With respect to the package delivered to Jacobsen from Temple, Connally testified that he told his secretary to get a transcript of his testimony before the Watergate Committee to Temple. He said he did not know whether the transcript of his appearance in our offices was enclosed and that in a conversation with Temple he told Temple he was going to send it to Temple because if the Ervin Committee reconvened, he wanted to get Christian's and Temple's opinions on it. He also said that he told Temple he could show it to Jacobsen. Temple, of course, will testify that he only glanced at it, that he gave it to Jacobsen the next business day without retaining a copy for himself or Christian. Connally also stated that he did not make copies of the transcript for anyone else as far as he knew.

Finally, Connally testified that he first heard about Jacobsen's indictment by the Watergate Grand Jury through the public domain. He was asked whether Christian or Temple told him they had learned about it from a source other than the public domain and he said no. He was asked whether either of them told him they had spoken to Jacobsen and he said no. He was specifically asked whether Temple told him "before the indictment was returned that he had spoken with Mr. Jacobsen and that Mr. Jacobsen had informed him that he, Mr. Jacobsen, was to be indicted?"

"No, I certainly do not recall the conversation."

III. ANALYSIS

A. *Weaknesses*

[DELETION: Discussion and evaluation of the status of potential defendant. Includes personal information concerning the individual. The evaluation is the attorney's opinion of the vulnerability and potential weakness of a case.]

The answers to this are the usual answers since this problem is certainly not unique. In addition to the usual answers, there are two factual points which apply with particular emphasis to this argument.

First, Jacobsen in his original offer of proof to us did not attempt to "give us" other persons whom he had every reason to believe we would be interested in, e.g., Nelson or Mehren.

[DELETION: Identifies individuals who testified before grand jury and includes the attorney's evaluation of the testimony.]

Therefore, if Jacobsen were in fact to be fabricating it would be only logical for him to start with a story that might allow him to try to salvage his position in Texas and his law license and liberty.

Second, it is clear from what Jacobsen has told us and what we know, that Jacobsen had an opportunity to embellish which he did not take. Jacobsen could have said that Connally insisted on new money as he said he told Nelson before the second price support decision. Jacobsen could have told us he dangled the prospect of personal cash in front of Connally before the March 25 decision or at least that Connally inquired about it ahead of time. Certainly Jacobsen was alone enough with Connally in early 1971 to come up with a more dramatic and revealing story.

A second weakness in the case is that it is hard to believe that Connally would violate the law for a mere $10,000. There are three answers to this argument. First it is very unlikely that Connally in 1971 felt he was risking anything in dealing with Jacobsen whom he had known and worked closely with for almost one quarter of a century.

Second, since Jacobsen was dealing with Lilly, a mere professional acquaintance who was much further removed from Jacobsen than Jacobsen was from Connally, it was Jacobsen who is taking the real risk. It is inconceivable that Jacobsen should in his April 28 conversation with Lilly distort the truth in favor of criminality by asking for money for Connally personally when it was supposed to be for political candidates.

[DELETION: Identifies individuals who testified before grand jury and includes the attorney's evaluation of the testimony.]

As a practical matter, therefore, the person who was actually taking the risk at the time was Jacobsen. It is unlikely that he would do so unless he meant what he said when he took it.

Third, the argument that Connally would not violate the law for a mere $10,000 is weakened by the narrow factual issues of the case. The issue is not whether Jacobsen gave Connally the money or stole it, but rather whether Jacobsen gave Connally the money or offered it to him on June 25, 1971 and then after it was declined, again in August 1972. Just five weeks before Jacobsen received the money in May of 1971, AMPI had reaffirmed its $2 million commitment to the Administration.

[DELETION: Identifies individuals who testified before grand jury and includes the attorney's evaluation of the testimony.]

In any event, Connally's statements on the March 23 tape alone reveal his knowledge of the dairy industry's potential largesse. Under these circumstances, it would be almost an insult to Connally to give him a mere $10,000 for contributions when AMPI was talking in terms of $2 million. Unlikely as it is that Connally would violate the law for a mere $10,000, it is equally unlikely that AMPI would lawfully deal in such small sums with him. The same analysis applies to 1972 events. SPACE contributed $25,000 to Democrats for Nixon. Just before the election in October, 1972, AMPI contributed another

$300,000 to the Republican Congressional effort. Certainly the offer of $10,000 to Connally at that time as a political contribution would have been equally unlikely.*

The third weakness in the case is probably the major weakness.

[DELETION: Summaries of or references to grand jury testimony.]

Connally's logs show that on October 14 he met with Jacobsen for one-half hour.

[DELETION: Summaries of or references to grand jury testimony.]

Jacobsen's records for box #998 reveal a 12:30 entry on November 10, 1971, Jacobsen's first entry into that box since February 3, 1970 and only his second entry since April of 1968. On December 15, 1971, Jacobsen came to Washington. Records for box #998 show a Jacobsen entry on 2:20, December 14, 1971.

From the beginning, Jacobsen has said that he did not recall receiving an additional $5,000 from Lilly. After intensive questioning about it and after he reviewed the safe deposit records referred to above, Jacobsen said that he must have gotten it from Lilly and that if he asked Lilly for the money for Connally, then he must have given it to Connally. He says, however, that he has no recollection of giving it to Connally and that his only recollection is traveling to Washington with an additional $5,000.

Connally has little room in which to argue that Jacobsen kept the $5,000 for himself. This is so because at the time he obtained the $5,000, Jacobsen was still a multimillionaire with no storm clouds on the horizon. Since Connally has acknowledged that Jacobsen offered him $10,000 in August of 1972, two months after Jacobsen was found bankrupt, it makes little sense to argue that he would steal $5,000 while still a very wealthy man.

What the defense can argue is that it makes no sense to cover up for $10,000 when $15,000 was paid. Our answer is that Jacobsen simply forgot about the third payment, and it was only after he went over his safe deposit records

[DELETION: Summaries of or references to grand jury testimony.]

that he remembered, to the extent that he does.

---

*As of August 1972, neither Jacobsen's law firm nor Jacobsen himself was attorney for AMPI.

B. *Strengths*

The strengths of the case fall into two distinct categories, areas where Connally is directly contradicted by other witnesses, and areas where the logic of the fact points to acceptance of Jacobsen's version of the events. Each will be discussed in turn.

(1) *Contradictions of Connally*

[DELETION: Segments identify individuals who testified before the grand jury and include the attorney's evaluation of the testimony.]

[DELETION: Segments identify individuals who testified before the grand jury and include the attorney's evaluation of the testimony.]

[DELETION: Segments identify individuals who testified before the grand jury and include the attorney's evaluation of the testimony.]

A third instance in which the evidence contradicts Connally involves Connally's testimony that he called Jacobsen back after returning to the Sheraton-Crest on October 26. The business records of the hotel reveal no local calls charged to Connally's phone and the evidence as to the manner in which charges for such calls were posted makes it very clear that Connally did in fact make no local calls. His only out is that after receiving Jacobsen's message, he called Jacobsen back from a phone booth in the lobby. The picture of John Connally hunching in a telephone booth, putting a dime in the machine to call Jacobsen in a hotel lobby before returning to his room is not a likely one.

[DELETION: Segments identify individuals who testified before the grand jury and include the attorney's evaluation of the testimony.]

[DELETION: Segments identify individuals who testified before the grand jury and include the attorney's evaluation of the testimony.]*

Connally is also contradicted by Connally on three matters. First, his knowledge of the source of the $10,000 is in conflict. Second, on November 14, Connally testified it had been a long time, perhaps fall of 1972, since he had last discussed the $10,000 with Jacobsen. On April 11, Connally testified that in an "extraordinary conversation" they talked about it less than three weeks before the November 14 testimony—on October 26. Third and equally important, on Novem-

---

*Connally said the meeting on October 26 was around 5:30 p.m.

ber 14, Connally testified that the October 29 business meeting was his only contact with Jacobsen in the three to four weeks preceding his testimony. This testimony was also markedly contradicted by the April 11 version which referred to the October 26 meeting at the Sheraton-Crest. The significance of these latter two contradictions is self-evident. There can be little room for doubt that Connally deliberately intended to mislead the grand jury on the nature and frequency of his pre-testimony contacts with Jacobsen.*

(2) *Circumstantial Evidence*

In a case of this sort, many inferences can be found supporting one witness' version or another. They vary in strength and we do not in this memorandum attempt to point out every inference, however weighty or unweighty, that can be drawn. For example, that all major contacts between Connally and Jacobsen from the events of October 24-29 until Jacobsen's indictment here were initiated by Connally— the November 9 debriefing, the George Christian meeting, the transfer of Connally's Washington testimonies—*in all cases through intermediaries,* vaguely supports the notion that Connally was up to something improper. We will not, however, attempt to cull from the facts of this case other arguments of comparable force. Rather, we point to what we believe are the four most significant areas of circumstantial support for Jacobsen's version.

a. The money.

First, it should be noted that the money found in Jacobsen's box on November 27, 1973, was too recent to have been put there pursuant to the story

[DELETION: References to grand jury testimony.]

and, where appropriate, agreed to by Connally. We will prove at trial that sixteen of the bills found in Jacobsen's box on November 27, 1973 were not in circulation on June 1, 1971, three and one-half weeks after Lilly gave Jacobsen the money, that three were not in circulation as late as October and that one did not make it until March, 1973.

[DELETION: Summaries of grand jury testimony which include the attorney's evaluation of the testimony.]

Third, the absence of Connally bills and presence of 49 Shultz bills

---

*It should also be noted that toll records of the 13 minutes October 24, 1973 call and the three-minute November 13, 1973 call ... [DELETION: References to grand jury testimony.] and conflict with Connally's testimony.

[DELETION: Summaries of grand jury testimony and include the atorney's evaluation of the testimony.]

Finally, that all of the "October 29" bills were in circulation as of that date is also significant, especially with respect to the 49 Shultz bills. While it is not surprising that the "pre-Connally" bills were in circulation as of October 29, the period between October 29 and February 28 being a small fraction of the time these bills had to be distributed to the public, this is not so with respect to the Shultz bills. Bureau of Engraving and Printing records reveal that the Shultz bills in general were first circulated to the public in very late 1972, early 1973. The time between October 29 and Jacobsen's decision to co-operate is approximately one quarter of the period. We have traced most of the October 29 currency including the Shultz bills as far as documentary evidence will take us and that pursuit reveals so far that all of the bills, including the Shultz bills, were in circulation as of October 29, 1973.

[DELETION: Summaries of grand jury testimony which include the attorney's evaluation of the testimony.]

Moreover, since these bills had been in circulation for such a short period, there is every chance that a post October 29 Shultz bill would appear among the 49. The absence of such a bill points to an origin of the $10,000 on or before October 29, long before Jacobsen knew of his problems.

b. Washington Safe Deposit Box Records/September 23-24 Logs

It is, of course, true that the safe deposit box entry records at American Security and Trust Company do not tell us what took place in Connally's office. We start, however, with the suspicious fact that the only two entries are on each occasion within one hour of a Jacobsen/Connally meeting. Since the first entry was made when the box was opened, the only plausible inference is that something was put into the box on that date, May 14, 1971. Since the box was found to be empty after the second entry, it is also compelling that whatever was put into the box on May 14 was taken out on the date of the second entry—September 24, 1971.

Connally's logs show that Jacobsen and Connally met for 45 minutes on September 23, 1971, a meeting that was arranged in advance as evidenced by the entry in Connally's appointment calendar. The next day Connally was very busy. After meetings at the White House which lasted until 10:30, he had an appointment with the Senate Foreign Relations Committee at 11:00. Nevertheless, he had time to meet with Jacobsen from 10:35 to 10:45, despite having spoken to him 45 minutes the day before.

[DELETION: Segments identify individuals who testified before the grand jury and include the attorney's evaluation of the testimony.]

Given their extensive meeting of the day before, their telephone contacts and the hectic day Connally had on September 24, there is considerable inferential support for the idea that the meeting was set up in order for Jacobsen to give something to Connally pursuant to an arrangement of the day before and that in fact he did. This inference becomes all the stronger when taken in conjunction with the emptying of the safe deposit box shortly before the meeting.

c. Texas Safe Deposit Entries.

[DELETION: Segments identify individuals who testified before the grand jury and include the attorney's evaluation of the testimony.]

Ragsdale Aviation records show that Jacobsen returned to Austin at 1:30 p.m. Records of the Citizens' National Bank show Jacobsen entering box 865 at 2:00 p.m. Of course, there is no entry into his box on November 26, 1973, the first day after the George Christian meeting.

[DELETION: Segments identify individuals who testified before the grand jury and include the attorney's evaluation of the testimony.]

[DELETION: Segments identify individuals who testified before the grand jury and include the attorney's evaluation of the testimony.]

Obviously, if Jacobsen, on his own, thought he should change the money, he could have done so at any time between November 2 and November 26, a period of fifteen business days. He chose, however, to enter on November 26, just a day after the undisputed evidence shows that he had seen Connally alone. The inference arising from this pattern of seeing Connally alone and then going to the box is plain. Connally gave something to Jacobsen.*

d. The October 29 Meeting

The October 29 meeting is obviously critical in the case. Connally

---

*The imminence of the inventory cannot be considered a factor in leading to the November 26 entry since the F.B.I. agents have stated that they did not inform Jacobsen until the morning of November 27 that they wanted to inventory the contents of the box.

has, from the beginning, attempted to insulate it from the $10,000 and has only worsened his position as a result.

[DELETION: Segments identify individuals who testified before the grand jury and include the attorney's evaluation of the testimony.]

Connally testified that Wortham had asked him about the bank matter on October 10, almost three weeks before and that because Wortham was such an important client he wanted the matter taken care of before his two week absence from Houston, which was to begin on October 29.

This, of course, does not explain why Connally had to see Jacobsen rather than talk to him over the phone especially since Connally said he did not give Jacobsen any documents and wanted to save time. This then is the first weakness in Connally's position. Connally disputes this, saying it was because he thought he could save time, that he wanted to see Jacobsen and this was "the" reason for his asking Jacobsen to come. This answer, which is not really an answer, only gets him into hot water on another matter.

It is not the illogic of not telephoning Jacobsen alone, however, which makes Connally's story unlikely.

[DELETION: Segments identify individuals who testified before the grand jury and include the attorney's evaluation of the testimony.]

Furthermore, if Connally really did want the matter taken care of before he left for Europe because of the past delay, it would follow that he would call Jacobsen after his return to find out where it stood. Connally did not do so.

[DELETION: Segments identify individuals who testified before the grand jury and include the attorney's evaluation of the testimony.]

A second powerful argument supporting the idea that the Gus Wortham application was simply a ploy to justify the meeting is Connally's testimony on April 11 that since he had heard and observed that Jacobsen was not the old Jacobsen, one of the reasons he wanted to see Jacobsen on October 29 was to be sure that he could handle the sensitive matter. In short, Connally wanted to judge his demeanor and not just hear his voice. While this may answer the troubling question as to why he did not just use the telephone, it serves to create two additional problems for Connally.

First and least important, it is inconsistent with Connally's earlier testimony that "the" reason that he wanted to see Jacobsen was to save time. Second, it is at total variance with Connally's own acknowledgment that he saw Jacobsen just three days earlier on October 26 and in fact was sufficiently perceptive of Jacobsen so that he sensed that when Jacobsen offered to lie, he was just feeling Connally out. In short, this "reason" which Connally testified to is a demonstrably false exculpatory statement.

C. *Conclusions*

The third payment problem is obviously the weakness in the case. It is not, however, as if Connally were a prospective defendant without his own very serious problems. Connally does have such problems, many of them in critical areas of the case. He has testified as to two distinct versions of the events preceding his November grand jury testimony. If he persists in the second version, he will be contradicted by a simple but strong witness—

[DELETION: Segments identify individuals who testified before the grand jury.]

If he changes his story to avoid collision with

[DELETION: Segments identify individuals who testified before the grand jury.]

he will be presenting another version—a third. Three different and self-contradictory statements under oath about material matters is a far from ideal position for a defendant.

It should also be noted that what starts off as basically a one-on-one case is really not that at all. There are several items of evidence

[DELETION: Segments identify individuals who testified before the grand jury.]

which contradict Connally on material matters. Furthermore, Connally's own contradictions and false exculpatory statements add to the weight of the case against him.

[DELETION: Identifies an individual who testified before the grand jury and includes the attorney's evaluation of the testimony.]

IV. RECOMMENDATIONS

We recommend submission of the attached indictment to the grand jury with a request that it be voted on.

COUNT 1: Jacobsen will plead guilty to this count.

COUNT 2 and 3: These counts charge Connally with acceptance of an illegal gratuity, once on May 14, 1971, and once on September

24, 1971. The only conceivable legal issue that could arise on these counts is that Connally's assistance on the price support issue was not an official act as defined by § 201. The memorandum previously submitted deals with and disposes of this issue.

The only conceivable factual argument in terms of sufficiency that could arise rests on Connally's failure explicitly to say to Jacobsen on April 28, 1971, "get me some money because of my help." The facts we can prove on this issue are as follows:

1) Jacobsen asked Connally to help on price supports.
2) Connally agreed to help.
3) Connally did in fact help, enormously.
4) Connally told Jacobsen he had helped.
5) Connally asked Jacobsen if he would get some money from AMPI whom he heard was being generous with its money and get it for him instead of all those politicians.
6) Jacobsen immediately thereafter asked Lilly if he would get $10,000 together for Connally who had been so helpful on price supports.

Under these circumstances, the jury could certainly conclude, indeed would almost have to conclude, that when Connally asked for the money, something he did on no other occasion, he expected payment because of the assistance he had given.*

Case law under section 201(g) supports the sufficiency of the evidence. In *United States* v. *Brewster*, 408 U.S. 501, 527 (1972) the Supreme Court stated that "to sustain a conviction it is necessary to show that appellee solicited, received . . . money with knowledge that the donor was paying him compensation for an official act . . .: evidence of the Member's knowledge of the alleged briber's illicit reasons for paying the money is sufficient to carry the case to the jury." The jury could certainly infer from the facts we can prove that Connally knew that he would be paid because he had helped: there was no other reason to expect payment.

Other cases are also useful. In *United States* v. *Kenner*, 354 F2d (2d Cir. 1965), *cert. denied*, 383 U.S. 958 (1966), Kenner, who was charged under section 201(f), asked I.R.S. auditors to pull returns out of the ordinary review process, suggested disallowances, and after his suggestions were followed, paid the auditors. In response to his claim that the evidence was insufficient to show that payments were for an official act, the Court said: "Far from being insufficient, the evidence leaves no room for any other conclusion than that the payments were made for the pretended audits." *Id.* at 784. Similarly,

---

*[DELETION: Conclusion of the prosecuting attorney concerning the decision to prosecute or not to prosecute.]

in *United States* v. *Roberts,* 408 F.2d 300 (2d Cir. 1969) a conviction under section 201(g) was upheld where the evidence showed performance of favors by Roberts and that money was put into a drawer "pointedly left open by Roberts."

COUNT 4: This count charges Connally with conspiracy with Jacobsen to obstruct justice and make false declarations. The evidence clearly requires such a charge. The only issue is whether the count should be narrowed to exclude conspiracy to make false declarations and obstruct justice before all bodies other than the August 13, 1973 grand jury. We feel that since, in fact, Connally and Jacobsen were at least as concerned with the Ervin Committee, a true picture of the charge is one that includes the conspiracy to lie and obstruct before the Ervin Committee. Furthermore, with the broader charge there would be no doubt at all about our ability to use portions of the Erwin Committee testimony. Connally would be able to do so in any event.

COUNT 5: This count charges Connally with perjury, pursuant to the conspiracy. It stands or falls with the rest of the evidence. There are no legal issues presently apparent. Although the unlawful agreement was made in Texas, venue on conspiracy and perjury exists in the District of Columbia since the overt acts of testifying took place here.

COUNT 6: This count charges Connally with giving false testimony when he testified that he had not spoken with Jacobsen for a long time about the $10,000 on November 14. Clearly the evidence warrants the charge. The only issue is whether Connally's April 11 testimony that he spoke with Jacobsen about the $10,000 on October 26 constitutes a recantation, an affirmative defense provided by section 1623(d).

Connally never acknowledged the falsity of this November 14 testimony.* Such an acknowledgment has been found to be a precondition to the invocation of section 1623(d). *United States* v. *Crandall,* 363 F. Supp. 648, 654 (W.D. Pa. 1973). Moreover, since Connally never acknowledged the falsity of the November 14 testi-

---

*When questioned on April 11, his attention was not directed to the part of his November 14 testimony in which he said it was a long time since he had discussed the $10,000 with Jacobsen, although it was directed to the general denial of contacts with Jacobsen.

A recent Third Circuit case has suggested that before a witness is prosecuted under section 1623 there may be a duty to inform him of the recantation defense. *United States* v. *Lardier,* (3rd Cir., 5/14/74) in Crim LWR 218. This brief summary of the case suggests that the defendant was not represented by counsel at the time of his appearance. Connally, of course, was represented by counsel. We have requested a copy of the full opinion.

mony, all we have is inconsistent testimony. Since the same Congress which passed section 1623(d) made the giving of inconsistent testimony a crime, Section 1623 (c), it would certainly not have considered an inconsistent version a recantation. See *United States* v. *Krogh*, 366 F. Supp. 1255, 1256, (D.D.C. 1973). ("His Ellsberg break-in affidavit given the prosecutor did not admit the prior falsehood but was merely in conflict with his prior statement.")*

Assuming arguendo that section 1623(d) does apply, a brief discussion of its origins is in order. When, in the Organized Crime Control Act of 1969, Congress supplemented the strictures against perjury provided by Section 1621 by allowing conviction for knowingly making two or more necessarily inconsistent statements, it also codified the defense of recantation, modeling the statute after New York Penal Code Section 210.25.**

The recantation provision of Section 1623 is found in subsection (d) and reads as follows:

(d) Where, in the same continuous court or grand jury proceeding in which a declaration is made, the person making the declaration admits such declaration to be false, such admission shall bar prosecution under this section if, at the time the admission is made, the declaration has not substantially affected the proceeding, or it has not become manifest that such falsity has been or will be exposed.***

The legislative history of this section suggests that the recantation provision was added to reverse the prior federal law, and to afford the witness who admits the falsity of his earlier testimony an oppor-

---

*It is also a condition precedent to the applicability of section 1623(d) that the proceedings be the same. At oral argument in the Fielding break-in case, Judge Gesell indicated his doubts that, under comparable circumstances, the proceedings would be deemed the same.

**We have thoroughly researched the legislative history of the Organized Crime Control Act. Section 1623 was the least controversial section of the act; it received little comment, and what comment was addressed to it was focused upon the elimination of the two witness rule and the provision making inconsistent declarations punishable without proving the falsity of one.

***Prior to the enactment of section 1623(d), federal courts had viewed "recantation" as a common law defense and rejected it, stating that it was relevant, if at all, only on the issue of willfulness. *United States* v. *Lococo*, 450 F.2d 1196 (10th Cir. 1971) *cert. denied* 406 U.S. 245; *United States* v. *Kahn*, 340 F Supp. 485 affirmed 472 F.2d 284 (1973) *cert. den.* 411 U.S. 983 (S.D.N.Y., 1971) *United States* v. *Worchester*, 190 F. Supp. 548, 569 (D. Mass., 1960 Wyzansky, J)

tunity to correct his earlier testimony and avoid prosecution. *See, e.g.* House Report on S.30 and Related Proposals at p. 47; Speech by Senator McClellan (Floor manager) on January 21, 1970 at Congressional Record S.334. However, in so doing the Congress noted that Section 1623(d) was "modeled upon a New York penal statute." (N.Y. Penal Law Section 210.25) House Report, *supra* at 48. Cases decided under that provision of New York law are, therefore, relevant. Briefly stated those cases interpret the New York provision,* as permitting recantation only where the witness does so promptly before the body conducting the inquiry has been deceived or misled to the harm of its investigation, and when there is no reasonable likelihood that his perjury may become known to the authorities. *See People* v. *Ezaugi,* 2 NY 2d 439 (1957). (Policemen testified falsely before a grand jury about a recorded conversation with a narcotics pusher. After his testimony, he learned that his partner had testified truthfully, he then did also, stating he was upset and unsure of Department strictures during his first appearance. Held: conviction of perjury affirmed, no recantation.) Accord, *People* v. *Ashby,* 8 NY 2d 238 (1960).

There is a difference between Section 1623(d) and New York law which minimizes the burden on Connally if taken literally. Under New York law, if a defendant retracts, the retraction is a defense if it was made before it substantially affected the proceedings *and* was made before it became clear that falsity would be exposed. Under the wording of Section 1623(d), if taken literally, Connally would only have to show only one if these two occurred: either no substantial effect on the proceedings *or* that falsity had not become manifest. He certainly cannot show the latter and probably not the former.

As to the requirement that it be manifest that the falsity had not been exposed, New York law, as has been noted, speaks of a "reasonable likelihood" that perjury has become known to authorities. In the three days before Connally's April 11, 1974, testimony, the newspapers were replete with reports that Jacobsen's and Connally's story was coming unglued. On April 10, for example, Jack Anderson stated that "witnesses have also given the F.B.I. statements which contradict Connally" and that Anderson asked Connally for his comments.

As far as showing that the proceedings were substantially affected, had Connally testified on November 14 as he did on April 11, Jacob-

---

*Section 210.25 Perjury; defense
In any prosecution for perjury, it is an affirmative defense that the defendant retracted his false statement in the course of the proceeding in which it was made before such false statement substantially affected the proceeding and before it became manifest that its falsity was or would be exposed.

sen would have been recalled and the entire investigation might have taken a different course. Needless to say, this is entirely speculative, but Judge Gesell has stated that "Section 1623(d) does not contemplate a detailed inquiry into the thought processes of Grand Jurors", 366 F. Supp. at 1256 and further, has very plainly intimated that once the Grand Jury has acted, as it did here in indicting Jacobsen, it is too late to recant.

COUNT 7: This count charges Connally with giving false testimony when he said on November 14 that his only contact with Jacobsen in the three to four weeks preceding his indictment was the business contact of October 29. On this count the only conceivable problem is the recantation problem discussed above. The arguments against such a defense are even stronger here since we are in a position to prove that the new Connally version of prior contacts is also false. Furthermore, when Connally was made aware of his earlier testimony, he did not acknowledge its falsity, but tried to argue that the answer was correct within the context of what preceded it.

COUNT 8: This count charges Connally with perjury in saying that Jacobsen did not tell him he had been subpoenaed when they spoke on October 28 over the telephone. Since the testimony was given on April 11, there is no recantation issue. Connally will have to risk either being contradicted by hotel documents,

[DELETION: Identifies an individual who testified before the grand jury.]

or change his testimony once again. Under the more credible version of the facts, the October 28 telephone call will be the first Connally/Jacobsen conversation after Jacobsen was subpoenaed. Connally's refusal to acknowledge that Jacobsen told him about it then is willful because of Connally's repeated and invariably futile attempts to clothe the October 29 meeting with regularity.

WATERGATE SPECIAL   DEPARTMENT OF JUSTICE
PROSECUTION FORCE
MEMORANDUM
TO   : Files   DATE: 6/20/74
FROM   : Frank M. Tuerkheimer
SUBJECT   :

The second sentence of the June 18 Connally memorandum, paragraph 4 in page 5, reads "Lilly will testify and his notes show a long distance call on April 28 from Jacobsen asking for $10,000 'for John Connally's lock box' for his help on the price supports."

After checking the notes and talking to Lilly, it turns out that a more accurate statement is that Lilly will testify that Jacobsen called him long distance on April 28 and said that because Connally has delivered for us, we're gonna have to deliver to him. You think you can come across with 10?: which Lilly understood to mean $10,000. Lilly's notes most likely written on May 4, 1971, the date he gave the $10,000 to Jacobsen has the following notation next to "April 28, 1971": "J. Jacobsen called me requesting $10,000 cast for J. Connally, be delivered [sic] to Jacobsen for placing in Connally's safe deposit box, Citizens' National Bank, Austin, Texas."

Jacobsen's box #865 was initially taken out in the name of the Democratic National Committee. Funds raised for the Committee were kept in that box.

Index

291